Estimating

with Microsoft® Excel

Third Edition

Estimating

with Microsoft® Excel

Unlocking the Power for Home Builders

Third Edition

Jay Christofferson

NAHB BuilderBooks

Estimating with Microsoft Excel®, Third Edition

BuilderBooks, a Service of the National Association of Home Builders

Courtenay S. Brown	Director, Book Publishing
Natalie C. Holmes	Book Editor
Torrie L. Singletary	Production Editor
Circle Graphics	Cover Design
Laserwords	Composition
The P.A. Hutchison Company	Printing
Gerald M. Howard	NAHB President and CEO
Mark Pursell	NAHB Senior Vice President, Exhibitions, Marketing & Sales
Lakisha Campbell	NAHB Vice President, Publications & Affinity Programs

Disclaimer

This publication provides accurate information on the subject matter covered. The publisher is selling it with the understanding that the publisher is not providing legal, accounting, or other professional service. If you need legal advice or other expert assistance, obtain the services of a qualified professional experienced in the subject matter involved. Reference herein to any specific commercial product, process, or service by trade name, trademark, manufacturer, or otherwise does not necessarily constitute or imply its endorsement, recommendation, or favored status by the National Association of Home Builders. The views and opinions of the author expressed in this publication do not necessarily state or reflect those of the National Association of Home Builders, and they shall not be used to advertise or endorse a product.

Printed in the United States of America

13 12 11 10   1 2 3 4 5

ISBN -10: 0-86718-647-X
ISBN-13: 978-0-86718-647-5

Library of Congress Cataloging-in-Publication Data

Christofferson, Jay C.
  Estimating with Microsoft Excel / Jay Christofferson. — 3rd ed.
    p. cm.
  Includes index.
  1. Microsoft Excel (Computer file) 2. Electronic spreadsheets. 3. Business—Computer programs. I. Title.

  HF5548.4.M523C487 2009
  005.54—dc22

                                          2009028522

For further information, please contact:

National Association of Home Builders
1201 15th Street, NW
Washington, DC 20005-2800
800-223-2665
www.BuilderBooks.com.

Contents

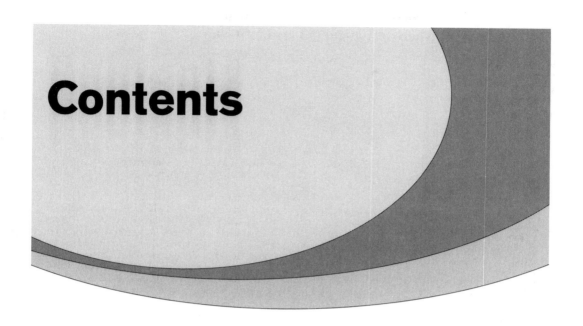

List of Figures   vii

List of Tables   xii

Foreword   xiii

About the Author   xv

Acknowledgments   xvii

Introduction   1

Chapter 1   The Basics of Spreadsheets and Beyond   5

Chapter 2   Improving Estimating by Using Spreadsheets   23

Chapter 3   Creating a Cost Breakdown Summary Sheet   35

Chapter 4   Eliminating Steps by Using Detail Sheets   47

Chapter 5   Saving Time with Hyperlinks   73

Chapter 6   Letting Formulas and Functions Do the Work   79

Chapter 7   Calculating Profit Margin   99

Chapter 8   Making Spreadsheets User Friendly with Form Controls   109

Chapter 9   Automating Spreadsheets   121

Notes   145

Resources   147

Index   149

Figures

Chapter 1

Figure 1.0. Office button, tab, and ribbon

Figure 1.1. Undo and Redo buttons

Figure 1.2. Excel® workbook

Figure 1.3. Sheet tab menu

Figure 1.4. Setting the default cursor movement

Figure 1.5. Selecting non-adjacent cells

Figure 1.6. Creating names from a selection

Figure 1.7. Selecting a named cell or range

Figure 1.8. Naming a formula

Figure 1.9. Creating a named cell

Figure 1.10. Pasting a formula name

Figure 1.11. Modifying a formula with Name Manager

Figure 1.12. Modifying a formula using the Edit Name box

Figure 1.13. Named constants dialog box

Figure 1.14. Using a constant in a formula

Figure 1.15. Naming cell ranges

Figure 1.16. Formula to return the value of intersecting ranges

Figure 1.17. Using Autofill to create data series

Figure 1.18. Formatting commands on the Home tab

Figure 1.19. Changing column widths

Chapter 2

Figure 2.1. Work breakdown items with variance

Figure 2.2. Cost control showing negative cost variance

Figure 2.3. Roofing item estimate in cost breakdown summary

Figure 2.4. Roofing detail estimate

Figure 2.5. Spreadsheet with roof slope options

Figure 2.6. Roofing detail for 8/12 slope

Figure 2.7. Basic detail sheet layout

Figure 2.8. Roofing material database

Chapter 3

Figure 3.1. Cost breakdown summary

Figure 3.2. AutoSum button

Figure 3.3. AutoSumming non-adjacent cells

Figure 3.4. Copying formulas by selecting cells

Figure 3.5. Copying formulas using fill

Figure 3.6. AutoSumming subtotals

Figure 3.7. Function Wizard

Figure 3.8. Insert Function dialog box

Figure 3.9. NOW function

Figure 3.10. Customizing Date Format

Figure 3.11. Spreadsheet with data displayed

Figure 3.12. Accounting number format and Increase or Decrease Decimal buttons

Figure 3.13. Paste menu

Figure 3.14. Paste Special dialog box

Figure 3.15. Paste Special: Multiply operation

Figure 3.16. Updated costs

Figure 3.17. Paste Special: Transpose

Figure 3.18. Sorting data

Figure 3.19. Data after sorting

Chapter 4

Figure 4.1. Detail sheet

Figure 4.2. Detail sheet with price extensions

Figure 4.3. Roofing database

Figure 4.4. Naming the roofing database

Figure 4.5. Accessing Data Validation from the Data tab

Figure 4.6. Data Validation: Allow List

Figure 4.7. Data Validation source

Figure 4.8. Data Validation: Drop-down list

Figure 4.9. Data Validation completed

Figure 4.10. Initiating VLOOKUP

Figure 4.11. VLOOKUP arguments

Figure 4.12. VLOOKUP results

Figure 4.13. Clearing error results with IF function

Figure 4.14. Finding units with VLOOKUP

Figure 4.15. Plan view of roof

Figure 4.16. Applying the Pythagorean theorem

Figure 4.17. Slope factor formula

Figure 4.18. Roofing squares calculation

Figure 4.19. Roofing database with formulas

Figure 4.20. Hip and common comparison

Figure 4.21. Hip and hip measurement

Figure 4.22. Roofing detail sheet

Figure 4.23. Roofing labor database

Figure 4.24. Roofing labor on detail sheet

Figure 4.25. The MATCH function in a formula

Figure 4.26. Roofing labor database and slopes

Figure 4.27. Changing cell formats

Figure 4.28. Changing cell protection

Figure 4.29. Protecting a worksheet

Chapter 5

Figure 5.1. Roofing detail sheet

Figure 5.2. Creating a link

Figure 5.3. Using a cell name in a link

Figure 5.4. Hyperlinking to the *RoofingTotal* cell

Figure 5.5. Hyperlinking to a cell reference

Figure 5.6. Hyperlinking to a Web page

Chapter 6

Figure 6.1. Flatwork detail and database

Figure 6.2. Named cells and cell ranges

Figure 6.3. Copying formulas using the Fill handle

Figure 6.4. Footing and concrete databases

Figure 6.5. Footing detail sheet

Figure 6.6. Formulas view of Rebar and Miscellaneous database

Figure 6.7. Valuation and Plan Check database

Figure 6.8. Permit table

Figure 6.9. Valuation calculations

Figure 6.10. Permit table with formulas

Figure 6.11. Permit detail sheet

Figure 6.12. Connection fees database

Figure 6.13. Connection fees detail

Figure 6.14. Adding comments to cells

Chapter 7

Figure 7.1. Construction loan detail

Figure 7.2. Construction S curve

Figure 7.3. Making a rectangle using the S curve

Figure 7.4. Using the rectangle to estimate construction costs

Figure 7.5. Breakdown of project costs

Figure 7.6. Breakdown of project costs–direct cost formula

Figure 7.7. Formulas to compute margin and markup

Figure 7.8. Sales price computation

Figure 7.9. Formulas used to determine sales price

Figure 7.10. Using iteration to determine sales price

Figure 7.11. Enabling iteration

Figure 7.12. Tracing tools

Chapter 8

Figure 8.1. Show Developer tab

Figure 8.2. Controls

Figure 8.3. Using check boxes

Figure 8.4. Anchoring a cell in a formula

Figure 8.5. Selecting materials using Option buttons

Figure 8.6. Control cell link used for Option button

Figure 8.7. Formula used with Option button

Figure 8.8. Grouping Option buttons

Figure 8.9. Scroll bar

Figure 8.10. Spinner

Figure 8.11. List Box

Figure 8.12. Making a List Box functional

Figure 8.13. Formulas used with a List Box

Figure 8.14. Completed List Box

Figure 8.15. Combo Box

Figure 8.16. Creating a Combo Box

Figure 8.17. Completed Combo Box

Figure 8.18. Making a selection with a Combo Box

Figure 8.19. Combo Box minimized after selection

Chapter 9

Figure 9.1. Relative reference

Figure 9.2. Record Macro button on the ribbon

Figure 9.3. Record Macro button on the status bar

Figure 9.4. Assigning a macro to the keyboard

Figure 9.5. Recording a macro

Figure 9.6. Return address macro in Visual Basic

Figure 9.7. Return address typed (top) and executed as macro (bottom)

Figure 9.8. Shapes menu

Figure 9.9. Assigning a macro to an object

Figure 9.10. Selecting the macro to attach to the object

Figure 9.11. Running a macro by clicking the object

Figure 9.12. Adding a custom macro button

Figure 9.13. Customizing the Quick Access Toolbar

Figure 9.14. Adding the macro to the Quick Access Toolbar

Figure 9.15. Modifying the macro icon

Figure 9.16. Editing the macro in Visual Basic

Figure 9.17. Final return address results

Figure 9.18. Creating a macro button

Figure 9.19. Placing a macro button on the worksheet

Figure 9.20. Inserting a User Form

Figure 9.21. Designing the User Form

Figure 9.22. Setting the *rowsource* property

Figure 9.23. Changing the Command Button caption to "OK"

Figure 9.24. Changing the Command Button caption to "Close"

Figure 9.25. Editing the text of a Command Button

Figure 9.26. User Form in Excel®

Figure 9.27. Displaying *Userform1*

Figure 9.28. Adding code to Command Buttons

Figure 9.29. Adding colors

Figure 9.30. Assign macro pop-up box

Figure 9.31. Setting the User Form properties

Figure 9.32. Testing the User Form

Figure 9.33. Adding code to place the cursor

Figure 9.34. Adding roofing items

Figure 9.35. Programming a User Form to respond to a double click

Figure 9.36. Indicating the end of a User Form

Figure 9.37. Adding rows to the User Form

Tables

Table 1.0. Excel® operators

Table 7.1. Monthly and accrued job costs

Table 7.2. Markups, margins, and multipliers

Foreword

Learning how to estimate the cost of jobs accurately and efficiently, in particular, is critical to building a profitable residential construction company. Computerized estimating is a standard tool in the construction industry. It is as important to your daily work as a working vehicle, a cell phone, and other tools of the trade. Microsoft Excel® is the most popular program used for estimating.

Following on the success of *Estimating with Microsoft Excel®, Second Edition* Jay Christofferson provides step-by-step instruction on how to use Microsoft Excel® 2007 to reduce the time you spend estimating construction jobs by letting your computer do most of the work. This book introduces the latest Excel® features and reviews the basics of using functions and formulas to ensure an accurate estimate every time and eliminate redundancy in your work.

Specifically, it explains how to use

- formulas for typical areas of construction estimating
- customizable *detail sheets*
- handy shortcuts to streamline estimating

After reading this book and practicing the examples, you will be able to complete estimates in minutes or hours, rather than days. You will be able to reach your profit goals by improving estimating accuracy, saving time, organizing jobs, and eliminating rework.

You entered the home building business because you love the work, but you will stay in business only if you can earn a reasonable profit. By improving your efficiency in estimating, you can build beautiful homes, make money doing it, and free up more time for doing the other important things you have to do.

—Leon Rogers
President
Construction Management Associates

About the Author

Jay Christofferson is a professor of Construction Management at Brigham Young University and the associate director of the School of Technology. He received a Ph.D. in Construction Management from Colorado State University. Jay is a licensed general contractor and has built hundreds of custom and production homes. He has been a consultant to many residential builders. A frequent speaker at the International Builders' Show, he is also the co-founder of Graduate Builder Seminars for more than 20 years. A commitment to developing computer solutions in management, communication, and estimating for construction companies led to his development of EstimatorPRO™ in 1998, an Excel®-based estimating program for builders and remodelers, which has been revised and updated five times. In 2007, he received the Outstanding Educator Award from the Home Builders' Institute, the workforce development arm of NAHB. He is an Eagle Scout and the scoutmaster for local Boy Scouts of America Troop 1074 in the Utah National Parks Council.

Acknowledgments

I would like to thank all those who have helped in any way and have given important input into the final product of this book, including Natalie Holmes, who has made beneficial edits and given important inputs, and to Courtenay Brown and the other staff members at NAHB's BuilderBooks. They have all assisted in many ways to make this work successful. I express my appreciation to the faculty members and students of the Construction Management Program at Brigham Young University for their input into the technical details of and methods included in this book.

Special thanks to my wife, Maxine, and to the other members of my family—Michael, David, Janae, Kelli, and Ryan—for their great support during the writing of this book.

Thanks also to Brian Blaylock, Anna Harb, John McVey, Chase Carlson and Michael Smith for their input and suggestions for the manuscript and to John Jones for his help with illustrations.

Introduction

Most people have Microsoft Excel® installed on their computers and they are increasingly proficient using the software. However, most users do not realize the full capacity and power of computerized spreadsheets in automating and simplifying a multitude of office tasks. Adapting to and becoming competent with computer skills, including using the full capability of Excel®, are analogous to using a power saw rather than a handsaw to build. The knowledge you will gain by reading this book and practicing the examples provided will help you transition to using the power saw.

One of my favorite cartoons shows a medieval general leading his army into battle. The army is arrayed for battle, with shields held high and spears in hand. The general stands next to an aide, who is trying to introduce a salesman to the general. The salesman is peddling a machine gun. The general's aide is trying to catch the attention of the preoccupied general but is waved off. The caption under the cartoon reads, "No! I can't be bothered with new technology . . . we have a battle to fight!"

As a builder, you understand the payoff for the time you spend to sharpen a saw or oil a tool. Likewise, knowing how to apply new technology to your construction business management practices is the only way to maintain your competitive edge. Without this knowledge, you may survive, but your business won't thrive. Although you may win the battle, you will lose the war. That is, although you may be able to break even or perhaps earn a small profit doing construction, you will not attain long-term profitability and growth.

This book presents the basics of using Excel® spreadsheets for new users and it offers tips for experienced users too. It is organized to help you scan for topics you are not familiar with and skip areas in which you are proficient.

The book will teach you how to create customized Excel® spreadsheets that will save you time and increase your professionalism. Following the book's guidelines, you will be able to create estimates that are both accurate and visually appealing.

You will learn how to set up and organize your spreadsheet estimate into a summary sheet with detail or work-up sheets backed by databases that store the cost information. More specifically, you will learn how to

- create and use formulas that will automatically calculate quantities
- use functions to look up information from databases and detail sheets
- link information
- navigate spreadsheets
- use macros to automate your work
- calculate profit and overhead accurately
- implement controls to add functionality to spreadsheets

To make the most of this book, you will need to have Microsoft Excel® 2007 installed on your computer. You can download a 60-day trial version of Microsoft Excel® 2007 online at www.microsoft.com. Simply search for Excel® 2007. This program will run with either the Windows Vista or Windows XP operating systems.

Each teaching module in this book includes step-by-step examples. Each Microsoft Excel® 2007 workbook on the companion CD corresponds to a chapter in the book and contains the completed worksheets illustrated in the chapter. These worksheets include both detail sheets and databases. The databases are identified by DB at the end of the worksheet names. You will notice that some workbooks are "macro enabled." (Their icon includes a yellow sheet with an exclamation point.) This means that they have an embedded computer code that will execute with a specific keyboard sequence or after a worksheet object is clicked. To allow the macros to run, you will need to save these workbooks using the save option "Excel® macro-Enabled Workbook."

To open the files

- Insert the CD into your CD-ROM drive.
- Open Excel®.
- Click the **Office** button, choose **Open,** and select the drive letter where the CD is located.
- Double click the file name, or select the file and click **Open.**
- **Save** a copy of the file to your hard drive.
- Open a blank worksheet, follow the instructions in the book, and compare your results with the completed worksheets on the CD.

Choosing Hardware

Your system hardware will determine how well Excel® performs. Upgrading the random-access memory (RAM) of your computer's hardware will usually increase the speed of your computer's processing ability. Newer computers generally are fast enough to accommodate the Excel® spreadsheet application.

In general, you will save yourself from a lot of frustration by upgrading your computer every four to five years. With up-to-date hardware, your software will run much faster and you will be able to work more effectively. You don't need the latest and most expensive computer system, but you should buy high-quality hardware. This will save you money in the long run by extending the useful life of your computer.

Conventions in This Book

- Instructions are listed step-by-step with Excel® commands in **bold** type.
- Formulas are in **bold** type.
- Notes of interest are highlighted using the following symbol .
- Multilevel menu commands are separated by a forward slash (/).

The Basics of Spreadsheets and Beyond

1

Computerized spreadsheets are essential to business management, and construction management is no exception. Microsoft Excel® is the most commonly used program for maintaining these spreadsheets. Even if you are proficient in Excel®, you need to invest time to practice using Excel® 2007. The program has a number of new capabilities, so this investment will be well worth the effort.

The Office button and tab-and-ribbon layout replace the traditional menus and toolbars. For the most part, the commands and features of earlier versions are there, but in different locations (fig. 1.0).

Finding Help

For specific questions about any Excel® function, button, or feature, use the Help button ? located in the upper-right corner (under the ✕ or Close button) of the ribbon. After clicking the Help button, type in a key word and choose from the topics displayed the one that addresses your question.

Figure 1.0. Office button, tab, and ribbon

Figure 1.1. Undo and Redo buttons

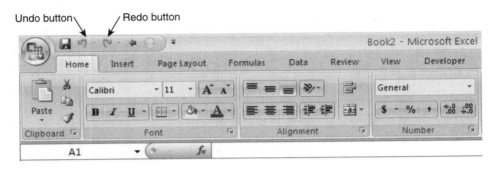

Undoing Mistakes

Don't be afraid to try something new. If Excel® doesn't behave as you expected it to, click the Undo button on the Quick Access toolbar (QAT) located to the right of the Office button. When you click the Undo button, the last instruction you performed will be undone. The second time you click it, the second to the last instruction you performed will be undone, and so forth. If you click the down-pointing arrow next to the Undo icon, Excel® will display a list of previous actions. By selecting one of the actions on the list, that action and those that followed it will be undone. If you decide that you really didn't want to undo, you can click the Redo button (fig. 1.1).

Using Workbooks

An Excel® *workbook* is the file in which you store and analyze data. Figure 1.2 shows a blank Excel® workbook. The Title bar displays the file name and application. The Office button allows you to set preferences for printing, saving, and other functions. The QAT allows you to display commonly used commands so you can see them no matter which tab on the ribbon is selected. You can also place icons for custom commands on the QAT, which will be discussed in later chapters. The Formula bar, which allows you to edit cell contents, is located above the column headings (A, B, C, etc.) and the Status bar, which alerts the user to the state of Excel®, is at the bottom of the page.

You can customize the Status bar by clicking the right mouse button (right click-ing) and choosing options from the menu such as page view, zoom, sum, average, and count.

Each worksheet contains 16,384 columns (A through XFD) and more than a million rows (1 through 1,048,576). The intersection of a column and a row is called a *cell*. Cells are storage bins that can hold numbers, text, formulas, and for-matting. Each cell has its own unique address. For example, the intersection of col-umn D and row 5 is cell D5. Although the data stored in cells can change, the cell addresses do not change.

Adding and Deleting Worksheets

Similar to a folder in a file cabinet, a workbook contains any number of related sheets. However, each worksheet consumes space and increases the size of the workbook so unless you need them, you are better off with fewer sheets in your

Figure 1.2. Excel® workbook

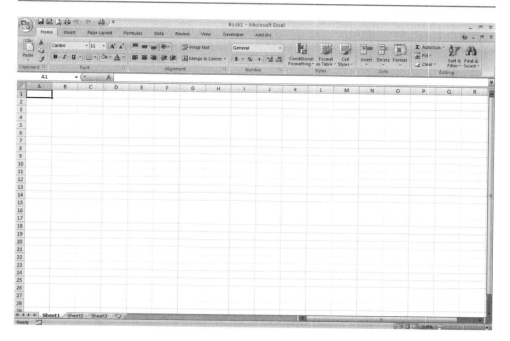

workbook. The default number of blank worksheets is 3. To change the default number of worksheets that open in a workbook

- Click **Office button/Excel® Options/Popular.**
- Select the desired number of sheets next to **Include this many sheets.**

You can add or delete worksheets from a workbook at any time. To add a worksheet

- Click the **Insert Worksheet** tab to the right of the worksheet tabs at the bottom of the workbook or right click one of the sheet tabs. If you use the first option, Excel® inserts the new sheet to the left of the Insert Worksheet tab. If you use the second option, Excel® inserts the new sheet to the left of the sheet that you right clicked.

To delete a worksheet

- Right click the tab of the sheet that you want to delete.
- Select **Delete** from the pop-up menu (fig. 1.3). A pop-up dialog box prompts you to either confirm the deletion or cancel.
- Select **OK** to delete the sheet.

Clicking the right mouse button will display a menu. Clicking the left mouse button will select an item or a cell.

Excel® automatically names worksheets with numbers: 1, 2, 3, etc. You can rename and move worksheets as follows:

- Double click the sheet's tab. The tab text is now in edit mode and can be changed.

Figure 1.3. Sheet tab menu

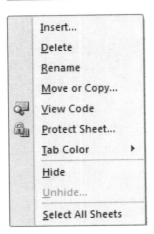

To move (reorder) a sheet, click and hold the mouse button while on the tab of the sheet you want to move and then drag the tab to the desired position.

Controlling the Cursor

When you type data or text and then press **Enter**, the cursor automatically drops down to the next cell. However, you can change this default movement so the cursor moves to the right, left, or up, or does not move after the **Enter** key is pressed. To change the default movement of the cursor

Figure 1.4. Setting the default cursor movement

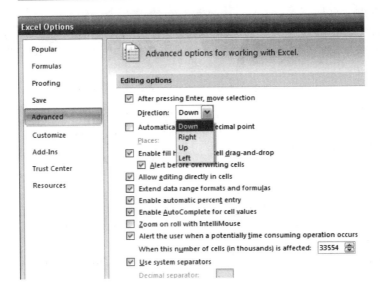

Figure 1.5. Selecting non-adjacent cells

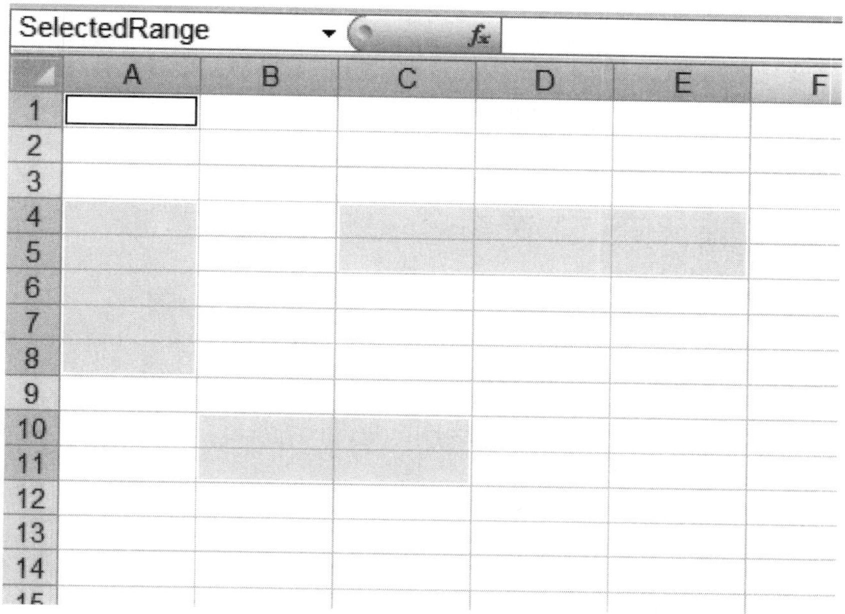

SelectedRange ▾		f_x			
A	B	C	D	E	F

(spreadsheet grid rows 1–15)

- Click **Office button/ Excel®️ Options/Advanced** and select the direction of movement (fig. 1.4).

I prefer to uncheck the "After pressing **Enter**, *move selection" box so the cursor does not move after the* **Enter** *key is pressed. This makes it easier to experiment with various "What If?" alternatives when preparing estimates.*

Selecting Cell Ranges

A group of cells is a *range.* You can change the default movement of the cursor to move within a specific range. To do this, select the range of cells and press **Enter.** To select a block of adjacent cells

- Select a cell in the corner of the block of cells and press **Shift.**
- Click a cell in the diagonal opposite corner of the desired block of cells.

All of the cells will be highlighted. When you press **Enter** within the block of cells, the cursor will move only within the block.

You also can select a group of non-adjacent cells.

- Press and hold the **Ctrl** key.
- Select each cell (fig. 1.5). Again, you will see the cells highlighted.
- Press **Enter** to move the cursor from one cell to the next within the selected range of cells.
- To move the cursor in the opposite direction, hold the **Shift** key while pressing **Enter.**

TABLE 1.0. Excel® Operators

Operator	Example	Returns the Value:
+ (addition)	= 2 + 5	7
− (subtraction)	= 15 - 5	10
* (multiplication)	= 4 * 3	12
/ (division)	= 20/10	2
^ (exponentiation)	= 3^2	3^2 or 9
% (percentage)	= 170 * 15%	25.5

Writing Formulas

All formulas begin with the equal sign (=). Formulas contain variables, littorals, and operators (table 1.0). Variables are values that may change. Littorals are values in formulas that do not vary from use to use (e.g., 3, 26.7, and 100.65). Operators are commands (add, subtract, multiply, divide, etc.) that indicate what to do with littorals or variables.

You could enter a formula into cell E2 to add the values of two variables as follows:

$$=B6 + C5$$

 The formula can be written with spaces, as shown, or without spaces (B6+C5). If the values stored in cells B6 and C5, respectively, were 5 and 72, the result of applying the formula would be 77. If, however, the variable in cell C5 were changed to 12, the result of applying the formula would be 17.

Naming Cells and Cell Ranges

By naming a cell or range of cells, you allow Excel® to use the name in formulas and as a reference in creating hyperlinks that allow you to easily move from one location to another within and between worksheets. For example, you could name a cell *MainFloorSF* to store the square footage of the main floor area of a home. Excel® then uses this name in formulas to determine the paint and drywall requirements, valuations for permits, framing labor, and so forth. Also, the named range, *MainFloorSF* could be used with the hyperlink feature to automatically move to the cell named *MainFloorSF* from another workbook location.

The range of cells shown in figure 1.5 is named *SelectedRange*.

To name a range

- Select the range of cells.
- Type a one-word name (no spaces) into the **Name:** box (to the left of the Formula bar).
- Press **Enter**.

This allows you to select the named range while working in any sheet in the workbook. Simply choose it from the list of names in the drop-down list in the Name box. You access this list by clicking the down arrow on the right side of the Name box.

Cells that have either row or column headings can be named automatically (fig. 1.6).

Figure 1.6. Creating names from a selection

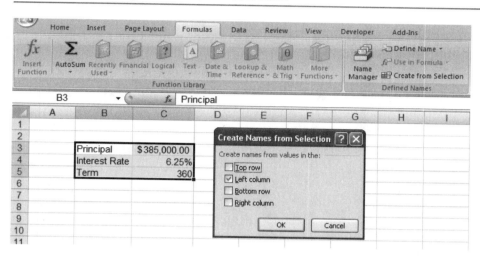

- Select the row heading cells and the cells to be named (**B3:C5**).
- Click the **Formulas** tab.
- Click **Create from Selection,** and verify that the labels for the cells you want named are in the column on the left.
- Click **OK.**
- Name the cells: **C3,** *Principal;* **C4,** *Interest Rate;* and **C5,** *Term.*

To check the labels, click cell C3 and verify that the name *Principal* is shown in the Name box.

Whenever the name *Principal* is selected from the drop-down list in the Name box, the cursor will move to C3 (fig. 1.7). Whenever you include the word Principal in a formula, Excel® will substitute the value in cell C3. For example, the formula = **Principal * Interest_Rate/12** would calculate the interest for one month.

Assume that the remaining principal on a loan balance is $180,000 and the annual percentage rate (APR) is 7.5%. Using the previous formula, Excel® would

Figure 1.7. Selecting a named cell or range

				f_x 385000	
Principal					
Interest_Rate			C		D
Principal					
Term					
3		Principal	$385,000.00		
4		Interest Rate	6.25%		
5		Term	360		
6					
7					

insert the appropriate variables and return the monthly interest as follows: $180,000 * .075/12 = $1,125.

Formatting Cells for Accounting and for Percentage

Cell C3, which holds the value 385000, was formatted to accounting style ($385,000.00) by selecting cell C3 and clicking $ on the Home tab. Cell C4 (.0625) is correctly shown as a percentage (6.25%) by selecting cell C4 and clicking % on the Home tab.

Rounding Up Values in Cells

The Round Up function in a formula looks like this:

ROUNDUP(number,num_digits)

The num_digits is optional. If you omit it or set it at zero, Excel® rounds up the number in question to the nearest integer as in the following two examples:

=ROUNDUP(27.374,) returns the value 28

=ROUNDUP(27.374,0) also returns the value 28

If num_digits is greater than 0, Excel® rounds up the number to the specified number of decimal places as in the following example:

=ROUNDUP(27.374,2) returns the value 27.38

If num_digits is −1, Excel® rounds up the number in the tens place as follows:

=ROUNDUP(27.374,-1) returns the value 30

You can apply rounding to construction in estimating cubic yards (CY) of concrete to the nearest ¼ CY as follows:

1. Multiply the number by 4.
2. Round up the result to the nearest integer.
3. Divide the result by 4.

Here's how the formula looks using the number from the previous examples:

=ROUNDUP(27.374*4,0)/4 returns the value 27.5

Naming Formulas

As with cells and ranges, you can assign names to formulas also. So, instead of manually typing or copying the same formula from place to place, you can simply copy the name of the formula.

Suppose you needed to estimate the area of a round concrete pad for a gazebo. You could create a formula to use anywhere in your workbook that calculates the area of a circle (πr^2). The Greek letter π (PI) represents a value of approximately 3.14. The letter r represents the radius of the circle. The radius of the circle is squared ($\wedge 2$).

The formula is as follows:

=3.14 * Radius^2

Figure 1.8. Naming a formula

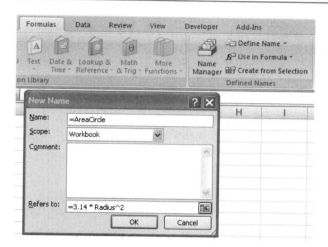

To name this formula

- Click **Define Name** in the **Formulas** tab. The New Name dialog box will appear (fig. 1.8).
- Enter the area of a circle formula in the **Refers to:** box.
- Enter the name *AreaCircle* in the **Name:** box.

To use this formula in a worksheet, you must assign a value to the radius variable. Type the word *Radius* in cell A1 and the number 2 in cell B1 (fig. 1.9).

Name the radius variable as follows:

- Select cells **A1** and **B1**.
- Click **Create from Selection** on the **Formulas** tab. The **Create Names** dialog box appears and asks if the name is in the left column. Excel® anticipates the correct answer and asks for verification. In this case, the name label is to the left of the cell where Excel® will store the radius value.
- Verify that the box next to **Left column** in the dialog box is checked.
- Click **OK**.

Figure 1.9. Creating a named cell

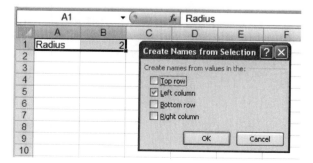

Figure 1.10. Pasting a formula name

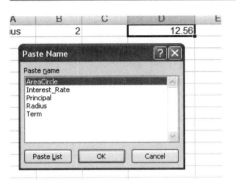

- **Paste** the formula's name in the cell where you want the area of the circle to display—in this case, cell D1, as shown in figure 1.10.

To display the Paste Name dialog box

- Press the **F3** key to display the list of available names.
- Select the appropriate name from the list (in this case *AreaCircle*).
- Click **OK**. (Alternatively, you could click **Use in Formula** in the **Formulas** tab and click **Paste Names.**)

Figure 1.10 shows the result of the formula in cell D1. The formula is displayed in the Formula bar. To test the formula, change the value of the variable in cell B1 and check the result.

Suppose you used this formula in several places in the workbook but then realized you had to change the formula to be more precise. Because you named it, you could change 3.14 to 3.141593 once instead of in every cell where the formula is used.

Change the formula for *AreaCircle* as follows:

- Click **Name Manager** in the **Formulas** tab.
- Change the formula in the **Refers to:** box (fig. 1.11).
- Click **Close** and then **Yes.**

Excel® updates all of the formula references for *AreaCircle* in the workbook. An alternative is to

- Double click *AreaCircle* after you click **Name Manager.**
- Change the formula in the **Refers to:** box in the **Edit Name** dialog box (fig. 1.12).
- Click **OK** and **Close.**

There are also two alternatives for typing the formula as follows:

$$=3.141593 * Radius^2$$

or

$$=PI() * Radius^2$$

Figure 1.11. Modifying a formula with Name Manager

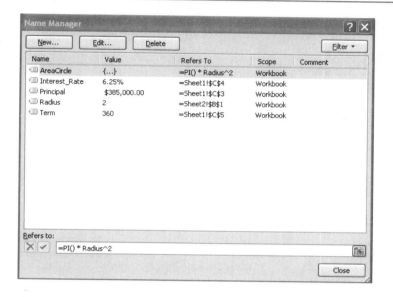

Figure 1.12. Modifying a formula using the Edit Name box

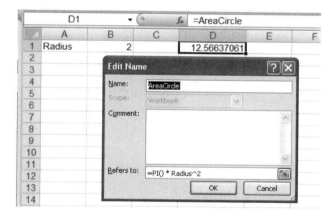

Naming Constants

Named constants can be as useful as named formulas as the following example, which calculates taxes (e.g., on materials purchased), illustrates. You estimate tax only on the materials that will be purchased, not all items. When you buy concrete, rebar, framing lumber, hardware, etc., you add a percentage (the tax rate) to the cost of the materials.

If you want to calculate the tax on concrete and your tax rate is 6.25%, or 0.0625, you can write a formula to take the subtotal of the concrete and multiply it by .0625. If the concrete subtotal is located in cell G28, the tax formula could be placed in G29. It looks like this:

$$=G28*.0625$$

Figure 1.13. Named constants dialog box

You could write a formula similar to this one for every item in your estimate that requires tax. You might use the tax formula 15 or 20 times throughout your estimate. However, if the tax rate increased to 6.75%, you would have to go back through each of the formulas and change the .0625 to .067—a time-consuming chore. A better option is to replace the tax rate with the constant *Tax* in the formula (fig. 1.13). So, instead of changing 15 or 20 tax formulas when the tax rate changes, you can change the value of the constant once, and all formulas that reference the constant will automatically update.

To create a tax constant

- Click **Define Name** in the **Formulas** tab.
- Enter the name of the constant (*Tax*) in the **Name:** box.
- Enter the value of the constant (=.0625) in the **Refers to:** box.

Figure 1.14 shows how Excel® uses the tax constant in the following formula, which calculates the tax amount:

Figure 1.14. Using a constant in a formula

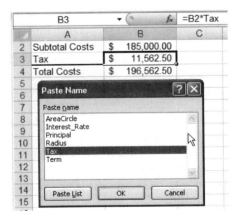

Figure 1.15. Naming cell ranges

	A	B	C	D	E	F	G	H
1	Community	January	February	March	April	May	June	Community Total
2	Oak Hills	$ 345,000	$ 276,900	$ 498,000	$ 530,000	$ 428,700	$ 405,000	$ 2,483,600
3	Riverbrooke	$ 219,000	$ 450,000	$ 789,900	$ 810,240	$ 689,000	$ 655,900	$ 3,614,040
4	Parkway	$ 543,200	$ 437,000	$ 635,400	$ 749,000	$ 587,000	$ 567,900	$ 3,519,500
5	Meadow	$ 168,000	$ 356,000	$ 430,400	$ 560,700	$ 395,000	$ 405,700	$ 2,315,800
6	Sunwood	$ 688,200	$ 495,000	$ 578,100	$ 605,000	$ 525,800	$ 480,000	$ 3,372,100
7	Monthly Total	$ 1,963,400	$ 2,014,900	$ 2,931,800	$ 3,254,940	$ 2,625,500	$ 2,514,500	$15,305,040
8								

Create Names from Selection [?][X]

Create names from values in the:

- ☑ Top row
- ☑ Left column
- ☐ Bottom row
- ☐ Right column

[OK]   [Cancel]

$$=B1 * Tax$$

Cell B1 is the subtotal cost of the material to be taxed. When the tax rate changes, the tax constant can be changed in the **Refers to:** dialog box and all formulas that use the tax constant will automatically update.

Finding Information in a Table

Using named ranges can help you search for information in tables. For example, you might track monthly sales for various communities in which you build. If you wanted to know the April sales for the Parkway community, you could create a formula to look up this information. One way to find specific data from a table is to use the *intersection operator*, which is the space between the two names, along with the named ranges.

First, name the rows and columns of a table.

- Select the table (including headers).
- Click **Create from Selection** in the **Formulas** tab (fig. 1.15). The **Create Names from Selection** dialog box prompts you to verify that the names for the columns are in the top row and the names for the rows are in the left column.
- Click **OK**.

Now, wherever you may be in the workbook, you can write a formula to instantly return the data you need from the table. For example, if you want to know the April sales for the Parkway community, enter the following formula:

$$=April\ Parkway$$

The value of the sales, $749,000, is the intersection of the *April* range and the *Parkway* range (fig. 1.16). The order of ranges in the formula does not matter in this case. If you wrote the formulas as **=Parkway April**, Excel® would return the same value.

Figure 1.16. Formula to return the value of intersecting ranges

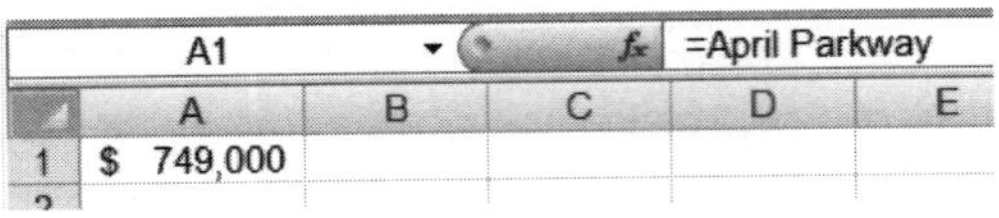

Entering Series of Data and Text

Although computers process information almost instantly, entering the information needed for the computer to do its work can be tedious. The row headings in figure 1.17 are months. Typing each month would be time consuming. However, the **Autofill** feature can simplify data entry. **Autofill** enables you to create series of numbers, dates, or other text by using the fill handle (the small black square in the lower right-hand corner of a selected cell or range of cells).

- Type the word *January* in A1.
- Drag the fill handle horizontally or vertically on the worksheet. Excel® completes the series of 12 months for you.

When you are working with numbers, dragging the fill handle will copy the number (fig. 1.17, column C). Holding the **Ctrl** key while dragging will produce an incremental change from the first cell (fig. 1.17, column D).

When you drag the fill handle down or to the right, the cells populate in increasing order. Dragging the fill handle up or to the left will enter data in decreasing order. To specify the desired increment of change (i.e., 0, 5, 10, 15, etc.)

- Type 0 in cell **E2** (fig. 1.17).

Figure 1.17. Using Autofill to create data series

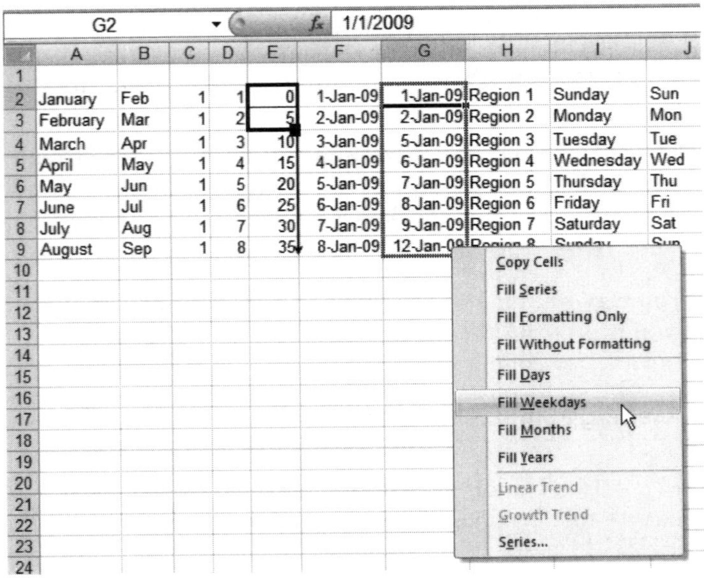

- Type 5 (the desired increment) in cell **E3**.
- Select both cells.
- Drag on the fill handle at the bottom right corner of the second cell to populate the desired cells.

Series are available for days of the week (columns I and J) and consecutive dates (column F). You also can create series with special instructions, as in column G.

- Type a date in G2.
- Select the cell.
- Right click and drag through the desired cells (G3 to G9).
- Release the right mouse button. Excel® displays a menu with additional functions.

Fill Weekdays will fill all days except weekends. This option is great for creating schedules. **Fill Formatting Only** copies the formatting of a cell but does not overwrite the cell's value or formula. **Fill Without Formatting** copies a cell or fills a series without changing the formatting in the target cells.

Formatting Cells

How your data looks is also important. Therefore, you should format your spreadsheets to look professional.

You can change the appearance, or formatting, of cells to fit specific needs. You can vary the size, features (boldface, italic, or underline), color, and font. Text can be centered, flush right, or flush left. You can shade cells using a variety of colors and you can format cells to display currency, dates, percentage, and many other standard and custom formats.

The easiest and fastest way to format a cell is to select it and click the appropriate command button on the **Home** tab of the ribbon (fig. 1.18). Another way to access the formatting commands is to right click the cell or range of cells to be formatted and choose **Format Cells** from the pop-up menu.

Use the icons on the Home tab to change a cell's appearance or the appearance of the text in the cell. The following options are available on the Home tab or on the pop-up menu:

- font
- font size
- bold
- italic
- underline

Figure 1.18. Formatting commands on the Home tab

- align left
- center
- align right
- merge cells and center
- wrap text
- currency style
- percentage style
- comma style
- increase decimal
- decrease decimal
- increase indent
- decrease indent
- borders
- fill color
- font color

Moving and Copying Cell Contents

You will often need to move or copy information from one location on your spreadsheet to another. The text, numbers, formulas, and formatting in cells can all be moved or copied to other locations.

To move a cell's (or range of cells') contents

- Select the cell or range.
- Click **Cut** ✄.
- Move your cursor to the new location.
- Click **Paste** 📋.

You also can move cells or ranges by using only the mouse:

- Select the cell or range.
- Move the cursor to the edge of the cell until the cursor changes to a four-pointed arrow.
- Click the edge of the cell, and while pressing the left mouse button, drag the cell to the new location.

To copy the contents of a cell (or range)

- Select the cell (or range) and click **Copy** 📑 on the **Home** tab of the ribbon.
- Choose a new location and click **Paste**.

There is an easier way to copy:

- Select the cell or range.
- Move the cursor to the edge of the selection until the cursor changes to an arrow.

- Press and hold the **Ctrl** key. Notice that a small **+** symbol appears next to the arrow cursor.
- Continue to press the left mouse button and the **Ctrl** key while dragging the cell or range to the new location.
- Release the mouse button.

A copy of the original cell (or cells) will be placed in the new location.

Changing Column Widths and Row Heights

The standard column width may be too narrow or too wide for the data or text stored in them. Both the width of columns and the height of rows can be changed to fit the data and text stored in them.

To change a column width

- Move the cursor to the right edge of the column heading. The cursor will change to a crosshair.
- Click and hold the mouse button and drag the column edge to the desired size.

Multiple columns can be resized to the same width. Just select across the column headings as follows:

- Click a column label (you will see a downward facing arrow) and drag across the desired number of columns.
- Drag any one of the column heading edges to the desired width. All selected columns will change to the same width.

To quickly resize the width of a single column or multiple columns to fit the size of the data they contain

- Select the column or columns and when the cursor changes to a crosshair double click at the right edge of any of the selected column headings (fig. 1.19). Excel® automatically resizes each column to fit the widest text or data in the respective column.

Row heights can be changed similarly.

- Select row headings instead of column headings.
- Double click one of the border lines between the row headings.

Figure 1.19. Changing column widths

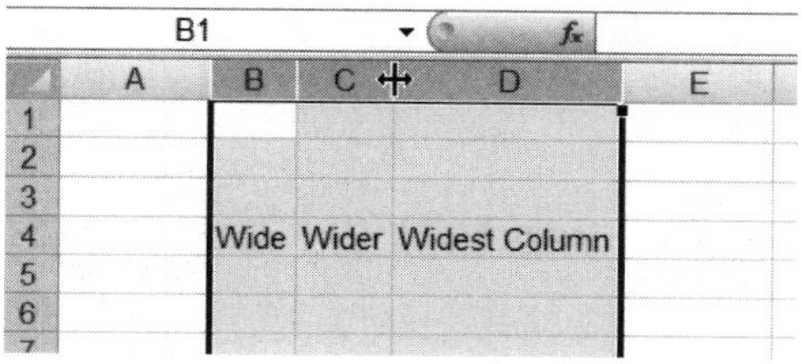

Practice These Basics

By practicing the basics in this chapter you will know more about Excel® than the vast majority of contractors. These basics will save you time and help you run your office more efficiently. You are on your way to unlocking the secrets and the power of Excel®.

Improving Estimating by Using Spreadsheets

Many builders and remodelers create detailed estimates manually, writing each line-item description and calculating quantities, perhaps by using a handheld calculator. They enter the unit or unit of measure next to the quantity. Next, they look up the unit price and write it next to the unit of measure. The item total—the extended price—is the product of the quantity and the unit price. Some builders and estimators have preprinted forms with lists of item descriptions and blank spaces where they enter quantities, units of measure, and unit prices. To calculate the item totals, they multiply the quantities by the corresponding unit prices, again using handheld calculators. Both of these manual methods are time consuming and prone to calculating errors.

Computerizing a preprinted form to automatically sum columns and calculate totals is a step forward in developing and delivering professional estimates faster. However, becoming proficient in using Excel® spreadsheets instead of the previous methods is like moving from U.S. mail to the cell phone.

Project Cost Breakdowns

Most contractors' estimates comprise a list of preconstruction and construction activities in their scheduled order (plans, permits, excavation, footings, foundation, etc.). Next to each activity is an estimated cost. The estimate is usually two to three pages and summarizes all construction costs. Builders, remodelers, and mortgage companies each have developed their own methods of organizing these construction cost breakdowns. A project cost breakdown summary allows builders to guard against omitting important items and to track and control costs. Mortgage companies and banks' mortgage departments also use project cost breakdown summaries when paying draws against construction loans.

Spreadsheet Estimating: A Better Way

In this chapter, you will develop a *cost breakdown summary sheet* that lists all construction items. Later in the book, you will learn how to create *detail sheets* to support the summary sheet. The summary sheet automatically retrieves summary costs from these worksheets, which contain the details of your cost estimating. In turn, these detail worksheets are supported by databases that store information you can access through the formulas and functions you enter into the detail worksheets. You can retrieve informational items, such as item descriptions, units of measure, and unit prices, for use in the detail sheets.

Typical Cost Breakdown Summary Sheets

A typical cost breakdown summary using the *course-of-construction* format (work packages listed in the order of the construction schedule) includes the following direct costs and markups:

Lot
Project overhead
    Building permit and fees
    Architectural fees
    Engineering fees
    Real estate sales fees
    Temporary utilities
    Construction loan
    Supervision
    Contingency
Hard costs
    Demolition
    Earthwork
    Footings
    Foundation
    Flatwork
    Miscellaneous steel
    Window wells
    Dampproofing
    Utility laterals
    Septic system
    Potable water well allowance
    Framing material
    Framing labor
    Entry doors
    Garage doors

Windows

Plumbing

Heating

Air conditioning

Electrical

Light fixture allowance

Roofing

Insulation

Drywall

Finish carpentry material

Finish carpentry labor

Painting

Tile/marble

Fireplace allowance

Floor coverings

Cabinets

Countertops

Appliances

Hardware & mirrors

Siding

Soffit & fascia

Gutter

Exterior railing

Foundation plaster

Cleanup

Landscaping

Profit and overhead

Company overhead (general & administrative)

Builder's margin (profit)

Appendix E of the *NAHB Chart of Accounts* (www.nahb.org/chart) contains an alternative course of construction organization under the *Direct Construction Costs, Subsidiary Ledger* that you can use to set up a cost breakdown summary sheet that will correspond to your accounting system. It includes the following accounts:

1010 Building permits

1020 HBA assessments

1030 Warranty fees

1110 Blueprints

1120 Surveys

1210 Lot clearing
1220 Fill dirt and material
1300 Demolition
1400 Temporary electric
1420 Individual wells
1430 Water service
1440 Septic system
1450 Sewer system
1460 Gas service
1470 Electric service
1480 Telephone service
1490 Other utility connections
2000–2999 Excavation and foundation
2000 Excavation and backfill
2010 Plumbing—ground
2100 Footings and foundation
2105 Rebar and reinforcing steel
2110 Concrete block
2120 Rough grading
2130 Window wells
2200 Waterproofing
2300 Termite protection
3000–3999 Rough structure
3100 Structural steel
3110 Lumber—1st package
3120 Lumber—2nd package
3130 Lumber—3rd package
3140 Trusses
3150 Miscellaneous lumber
3210 Framing labor—draw #1
3220 Framing labor—draw #2
3230 Framing labor—draw #3
3300 Windows
3350 Skylights
3400 Exterior siding
3410 Exterior trim labor
3500 Flatwork material
3550 Flatwork labor
3610 HVAC—rough

3720 Plumbing—rough

3810 Electrical—rough

3910 Gutters and downspouts

4000–4999 Full enclosure

4100 Roofing material

4150 Roofing labor

4200 Masonry material

4250 Masonry labor

4300 Exterior doors

4350 Garage door

4400 Insulation

4500 Fireplaces

4600 Painting—exterior

5000–5999 Finishing trades

5100 Drywall

5200 Interior trim material

5250 Interior trim labor

5300 Painting—interior

5400 Cabinets and vanities

5450 Countertops

5510 Ceramic tile

5520 Special flooring

5530 Vinyl

5540 Carpet

5610 Hardware

5620 Shower doors and mirrors

5630 Appliances

5700 HVAC—final

5710 Plumbing—final

5720 Electrical fixtures

5730 Electrical—final

5810 Wall coverings

5890 Special finishes

6000–6999 Completion and inspection

6100 Cleanup

6200 Final grade

6300 Driveways

6400 Patios, walks

6450 Decks

6490 Fences

6500 Ornamental iron

6600 Landscaping

6700 Pools

Some builders follow Construction Specification Institute's (CSI) *Master Format*[1]. The CSI Master Format divides construction costs into 50 major divisions. As in the *Chart of Accounts*, each major division is further divided into smaller subdivisions. Most commercial and industrial construction estimators use the CSI format to standardize the organization of estimates so that everyone who must refer to the estimate can easily find where it accounts for costs. Following is a list of CSI's divisions:

Procurement and contracting requirements group

    00 Procurement and contracting requirements

Specifications group

General requirements subgroup

    01 General requirements

Facility construction subgroup

    02 Existing conditions

    03 Concrete

    04 Masonry

    05 Metals

    06 Wood, plastics & composites

    07 Thermal & moisture

    08 Openings

    09 Finishes

    10 Specialties

    11 Equipment

    12 Furnishings

    13 Special construction

    14 Conveying equipment

15-19 Reserved

Facility services subgroup

    20 Reserved

    21 Fire suppression

    22 Plumbing

    23 Heating, ventilating, and air conditioning

    25 Integrated automation

    26 Electrical

    27 Communications

    28 Electronic safety and security

Site and infrastructure subgroup

    31 Earthwork

    32 Exterior improvements

    33 Utilities

    34 Transportation

    35 Waterway & marine

    36-39 Reserved

Process equipment subgroup

    40 Process integration

    41 Material processing and handling equipment

    42 Process heating, cooling, and drying equipment

    43 Process gas and liquid handling, purification, and storage equipment

    44 Pollution control equipment

    45 Industry-specific manufacturing equipment

    46-47 Reserved

    48 Electrical power generation

    49 Reserved

Any of these systems can be used as is or modified to fit your individual needs. No matter which method you use, having a system in place will allow you to efficiently create accurate estimates.

Cost Control Using the Cost Breakdown Summary

Construction cost breakdown summary sheets are typically two or three pages; however, some builders have created detailed cost summary sheets that are several pages. They are probably doing all of their estimating on the summary sheet and not using detail or work-up sheets. Each item on the cost breakdown summary sheet will have a budgeted cost assigned to it. As a builder or remodeler incurs costs for the work breakdown items, these costs are totaled and subtracted from the budgeted amount. The difference between estimated (budgeted) costs and the actual costs are called *cost variances*. Tracking these variances will help you control construction costs.

Figure 2.1 shows a partial construction cost breakdown summary. The estimated cost (budget) is in the third column. For footings, the estimated cost is $2,554.26. As invoices arrive, they are totaled in the appropriate cost category under one of the draw, or disbursement, columns.

Figure 2.1. Work breakdown items with variance

Code	Hard Costs	Est. Cost or Bid	Draw 1	Draw 2	Draw 3	Total Draws	Variance
	Demolition	$ -				$ -	$ -
	Earthwork	$ -				$ -	$ -
300	Footings	$ 2,554.26	$ 550.00	$ 1,854.00		$ 2,404.00	$ 150.26
	Foundation	$ -				$ -	$ -

Figure 2.2. Cost control showing negative cost variance

Code	Hard Costs	Est. Cost or Bid	Draw 1	Draw 2	Draw 3	Total Draws	Variance
	Demolition	$ -				$ -	$ -
	Earthwork	$ -				$ -	$ -
300	Footings	$ 2,554.26	$ 550.00	$ 1,854.00	$ 400.00	$ 2,804.00	$ (249.74)
	Foundation	$ -				$ -	$ -

All invoices dealing with the footings are coded with the number 300. In the first month of construction, the builder paid $550 to cover footing costs. The next month, the builder paid $1,854 in footing expenses. The variance, $150.26, is the amount left in the footing budget to cover any additional costs. If the project incurs no further footing expenses, this amount will be a surplus.

However, the following month the builder received a footing-related bill for $400 (fig. 2.2), turning the surplus into a deficit of $249.74. If all other variances are 0, this amount will be deducted from the profit margin. By scrutinizing variances throughout a project, especially in its early stages, you may be able to recoup excess costs incurred in one area by increasing efficiency in other parts of a project and, thus, earn a profit.

Because some builders and remodelers don't track variances during the course of construction, they do not know whether they have made a profit or incurred a loss until it is too late. By uncovering cost overruns early enough in a project, you can adjust material and labor costs to ensure a profit.

Detail Sheets

Each item on the cost breakdown summary has an estimated cost, or budgeted amount. These costs are derived using various methods. Some are an allowance, others are from trade contractor bids, and others are prices per square foot (SF) or per linear foot (LF). Some items are calculated using detailed quantity take-off and pricing methods. The detailed method is the most accurate way to estimate, but it is also the most time consuming. Using Excel® can shave hours off the time needed to create these detailed estimates.

If you are not keeping itemized detail sheets with your estimates, you can be sure that critical information will be lost or forgotten. This can be a real headache when you need to make changes. For example, figure 2.3 shows the estimated cost of roofing on the cost breakdown summary as $3,610.07. The builder in this case bid the job 7 weeks before roofing was to begin. When asked what the cost would be to change from a 6/12 sloped roof to an 8/12 sloped roof, the builder could only guess at the upgrade cost because he had not kept complete records of the roof details. To minimize the risk of guessing incorrectly, he had to create a new estimate.

Figure 2.3. Roofing item estimate in cost breakdown summary

Code	Hard Costs	Est. Cost or Bid	Draw 1	Draw 2	Draw 3	Total Draws	Variance
	Electrical	$ -				$ -	$ -
	Light Fixture Allowance	$ -				$ -	$ -
	Roofing	$ 3,610.07				$ -	$ 3,610.07
	Insulation	$ -				$ -	$ -
	Drywall	$ -				$ -	$ -

Figure 2.4. Roofing detail estimate

Roofing	Add Item		Choose Supplier	Supplier 1
Description	QTY	Unit	$/Unit	Sub-Total $
Asphalt Shingles - 30 yr. 3 Tab	34	Sq	$ 38.00	$ 1,292.00
Deliver & Stock shingles	37	Sq	$ 3.50	$ 129.50
Drip Edge-metal	30	Ea	$ 3.90	$ 117.00
30# Tar Paper (Asphalt-impregnated Felt)	17	2-Sq Roll	$ 10.95	$ 186.15
Starter Strip	1	Sq (T)	$ 38.00	$ 38.00
Hip & Ridge Cap	2	Sq (T)	$ 38.00	$ 76.00
Turtle-back Vents 10 x 14 (1 SF net area)	1	Ea	$ 15.00	$ 15.00
Step Flashing 3 x 4 x 7 (22' / Bndl.)	2	Bndl of 50	$ 12.95	$ 25.90
L Metal 4" x 4" x 10'	3	Ea	$ 4.79	$ 14.37
Plastic Caps 1/2 lb./sq	1	Pail	$ 34.95	$ 34.95
Coil Roofing Nails 1" - 1-1/4"	2	Ea	$ 37.95	$ 75.90
Labor /SQ for 6/12 slope	37	SQ	$ 40.00	$ 1,480.00
End				
Starter,Hip,Ridge Squares:	3		Labor Sub-Total:	$ 1,480.00
			Material Sub-Total:	$ 2,004.77
			Sales Tax: ☑	$ 125.30
			Total Cost:	$ 3,610.07

What Ifs

A detail sheet for roofing includes specifics such as slope, shingle type, and felt. Figure 2.4 shows the detail sheet that generated the roofing estimate shown in figure 2.3. The detailed estimate shows all materials and labor included in the roofing estimate and indicates that quantities were based on a 6/12-sloped roof.

Using this type of detail sheet, you could instantly recalculate the costs of changing the roof slope from 6/12 to 8/12. Figure 2.5 shows an example of a

Figure 2.5. Spreadsheet with roof slope options

Roofing Detail

Choose Estimating Method			Total Costs:
○ Lump Sum			$ 4,222.71
● Detail Estimate	⊕	$ 4,222.71	

	Slope 1	Slope 2	Slope 3	Total
Enter Plan SF of Roof Here:	2870			2870
Enter Rise / 12" Run:	8	6	6	
Slope Factor:	1.202	1.118	1.118	
Waste Factor:	5%	5%	5%	
Additional Squares:				0
Squares:	36.33	0.00	0.00	36.33

	Starter,Hip,Ridge Squares	3.00
	Total Squares	39.33

Enter LF of Fascia:	280
Enter LF Starter Strip:	140
Enter LF of Ridge:	82

	Slope 1	Slope 2	Slope 3	
Enter Total of Hip Dimensions A	44			
Hip Factor	1.563	1.500	1.500	
Waste Factor	5%	5%	5%	
Total LF	73	0	0	73

Figure 2.6. Roofing detail for 8/12 slope

Roofing	Add Item		Choose Supplier	Supplier 1
Description	QTY	Unit	$/Unit	Sub-Total $
Asphalt Shingles - 30 yr. 3 Tab	36	Sq	$ 38.00	$ 1,380.67
Deliver & Stock shingles	39	Sq	$ 3.50	$ 137.66
Drip Edge-metal	30	Ea	$ 3.90	$ 117.00
30# Tar Paper (Asphalt-impregnated Felt)	19	2-Sq Roll	$ 10.95	$ 208.05
Starter Strip	1	Sq (T)	$ 38.00	$ 38.00
Hip & Ridge Cap	2	Sq (T)	$ 38.00	$ 76.00
Turtle-back Vents 10 x 14 (1 SF net area)	1	Ea	$ 15.00	$ 15.00
Step Flashing 3 x 4 x 7 (22' / Bndl.)	2	Bndl of 50	$ 12.95	$ 25.90
L Metal 4" x 4" x 10'	3	Ea	$ 4.79	$ 14.37
Plastic Caps 1/2 lb./sq	1	Pail	$ 34.95	$ 34.95
Coil Roofing Nails 1" - 1-1/4"	2	Ea	$ 37.95	$ 75.90
Labor /SQ for 8/12 slope	39	SQ	$ 50.00	$ 1,966.50

End				
Starter,Hip,Ridge Squares:	3		Labor Sub-Total:	$ 1,966.50
			Material Sub-Total:	$ 2,123.49
			Sales Tax: ☑	$ 132.72
			Total Cost:	$ 4,222.71

spreadsheet designed to calculate an estimate using either slope. Note that additional columns are provided for building with multiple roof slopes, each of which can be changed. Figure 2.6 shows labor and material quantities for the 8/12 roof slope. Instead of guessing or redoing the entire roofing estimate, you enter a rise of 8 into the spreadsheet, and Excel® recalculates the cost instantly to $4,222.71. The 8/12 slope will cost $612.64 more than the 6/12 slope.

Excel® spreadsheets allow you to evaluate what-if scenarios instantly and effortlessly. You can immediately answer questions with cost implications such as "What if we changed to tile shingles?" Using these spreadsheets, you can quickly and easily calculate the costs associated with change orders.

Repair Shop Phenomenon

Using automated spreadsheets to calculate prices for upgrades and change orders has another advantage. I call it the *repair shop phenomenon*. If you take your automobile to a repair shop, you may have preconceived notions about the length of time required to fix the car. You quickly multiply the time it should take by how much you think you should be paying per hour. But to your disappointment, the service manager goes to the computer and looks up the charges for your specific repair. No matter how much higher than your own estimate the actual price for the repair may be, you don't try to bargain for a lower price. Who can argue with the computer? The price is what it is, and everyone gets charged the same amount. Your clients will have the same feeling when you use computerized spreadsheets. They will know you are a professional with established costs, and they will be less likely to question your estimates.

Detail Sheet Format

To create a detail sheet, you enter the formulas that calculate each category included in an item's total cost, such as roofing, into individual cells. To generate

Figure 2.7. Basic detail sheet layout

	A	B	C	D	E
1					
2	**Description**	**Quantity**	**Unit**	**$/Unit**	**Total $**
3					
4					
5					
6					
7				**Total Cost:**	
8					

item totals, multiply the quantities by the cost per unit. Excel® calculates the total cost by summing all of the individual item totals (fig. 2.7).

Databases

Data such as units and unit prices that are stored in a database support each detail sheet. A *database* is a list or a table of information. A roofing material database may contain item descriptions, units of measure, unit prices, and other important information (fig. 2.8).

By using Excel® to create cost breakdown summary sheets, detail sheets, and databases, you will reduce estimating time, improve estimates, easily accommodate changes to your plans and projects, and increase your profit margin.

Figure 2.8. Roofing material database

	A	B	C	D	
1	**Roofing Material Database**				
2					
3	**Description**	**QTY**	**Unit**	**Unit Cost**	
4	Asphalt Shingles - 25 yr. 3 Tab		Sq	$	30.00
5	Asphalt Shingles - 30 yr. 3 Tab		Sq	$	38.00
6	Asphalt Shingles - 30 yr. Architectural		Sq	$	48.43
7	Cedar Shakes #1 medium handsplits		Sq	$	110.00
8	Cedar Shingles #1		Sq	$	135.00
9	Metal Roofing		Sq	$	90.00
10	Eagle Tile		Sq	$	105.00
11	----------------------------------				
12	Deliver & Stock shingles		Sq	$	3.50
13	Drip Edge-metal		Ea	$	3.90
14	15# Tar Paper (Asphalt-impregnated Felt)		4-Sq Roll	$	10.95
15	30# Tar Paper (Asphalt-impregnated Felt)		2-Sq Roll	$	10.95
16	Ice and Water Shield - 3' x 75'		2-Sq Roll	$	77.00
17	Starter Strip		Sq (T)	$	38.00
18	Hip & Ridge Cap		Sq (T)	$	38.00
19	Architectural Hip & Ridge Cap		Box	$	30.00
20	Ridge Vents		Lf	$	2.30
21	Turtle-back Vents 10 x 14 (1 SF net area)		Ea	$	15.00
22	Step Flashing 3 x 4 x 7 (22' / Bndl.)		Bndl of 50	$	12.95
23	18" x 10' Valley Flashing		Ea	$	5.50
24	L Metal 4" x 4" x 10'		Ea	$	4.79
25	Plastic Caps 1/2 lb./sq		Pail	$	34.95
26	Simplex Nails-1/2 lb./sq		Lb	$	0.85
27	Roofing Nails - 1-1/4" - 2 lb./sq		Lb	$	0.62
28	Coil Roofing Nails 1" - 1-1/4"		Ea	$	37.95
29	8d galvanized box nails		Lb	$	0.64
30	6d galvanized box nails		Lb	$	0.64
31	4d galvanized box nails		Lb	$	0.64

Creating a Cost Breakdown Summary Sheet

I once worked with a builder who had a good year in terms of sales volume, but he did not earn a profit. All year long, company officials made decisions based on the assumption that the company was earning profits on each job. The company estimated, sold, and built without ever comparing actual construction costs with estimated costs. The builder did not realize until tax time arrived that the company had not made money. By then, it was too late to try to recoup the losses. If this builder had compared actual costs with estimated costs as each home progressed, he would have been able to seal the profit leaks that destroyed his bottom line.

By creating cost breakdown summary sheets in Excel® you will automate your work, save time, and be able to control costs more easily. The cost breakdown summary sheet greatly simplifies estimating and increases productivity because it

- instantly reflects any changes made in the details of the quantity take-off sheets
- provides professional documents for mortgage companies, buyers, and your internal office needs
- serves as a checklist to ensure the estimate does not omit items
- helps builders and remodelers control costs and increase profitability by tracking cost variances

You can design your cost breakdown summary sheet according to your preferences. Figure 3.1 shows one possible design.

Figure 3.1. Cost breakdown summary

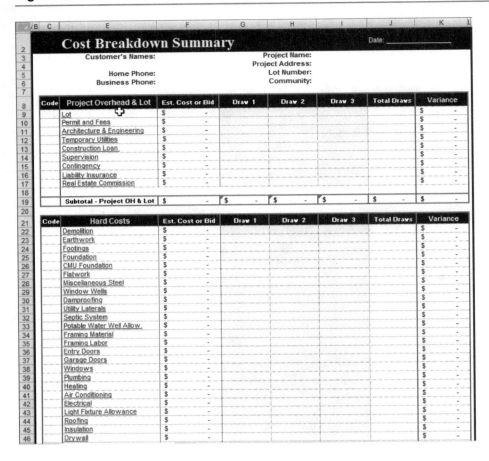

Code	Project Overhead & Lot	Est. Cost or Bid	Draw 1	Draw 2	Draw 3	Total Draws	Variance
	Lot	$ -				$ -	
	Permit and Fees	$ -				$ -	
	Architecture & Engineering	$ -				$ -	
	Temporary Utilities	$ -				$ -	
	Construction Loan	$ -				$ -	
	Supervision	$ -				$ -	
	Contingency	$ -				$ -	
	Liability Insurance	$ -				$ -	
	Real Estate Commission	$ -				$ -	
	Subtotal - Project OH & Lot	$ -	$ -	$ -	$ -	$ -	$ -

Code	Hard Costs	Est. Cost or Bid	Draw 1	Draw 2	Draw 3	Total Draws	Variance
	Demolition	$ -				$ -	
	Earthwork	$ -				$ -	
	Footings	$ -				$ -	
	Foundation	$ -				$ -	
	CMU Foundation	$ -				$ -	
	Flatwork	$ -				$ -	
	Miscellaneous Steel	$ -				$ -	
	Window Wells	$ -				$ -	
	Damproofing	$ -				$ -	
	Utility Laterals	$ -				$ -	
	Septic System	$ -				$ -	
	Potable Water Well Allow.	$ -				$ -	
	Framing Material	$ -				$ -	
	Framing Labor	$ -				$ -	
	Entry Doors	$ -				$ -	
	Garage Doors	$ -				$ -	
	Windows	$ -				$ -	
	Plumbing	$ -				$ -	
	Heating	$ -				$ -	
	Air Conditioning	$ -				$ -	
	Electrical	$ -				$ -	
	Light Fixture Allowance	$ -				$ -	
	Roofing	$ -				$ -	
	Insulation	$ -				$ -	
	Drywall	$ -				$ -	

Creating the Cost Breakdown Summary Sheet

The companion CD includes the worksheet *CostBreakdownSummary* for you to use as a reference in creating your own summary sheet. The Chapter 3-Book 1 workbook is ready for you to enter the formulas as explained in this chapter. The Chapter 3-Book 2 workbook is complete, with formulas included, so you can check your work.

First, create a new Excel® workbook and list all items you would normally want to estimate on a worksheet. You can use one of the sample lists from Chapter 2 or import a list from another source.

The first column holds the cost code numbers. You can use the *NAHB Chart of Accounts*, the CSI numbers, or your own system, which can be alphanumeric.

As bills arrive, you assign cost codes to them and enter the invoice amount into the appropriate draw (disbursement or actual cost) columns. To determine cost variances, subtract the dollar amount of the draws from the estimated costs. You can perform this task in Excel® by entering formulas into your cost breakdown summary sheet.

AutoSum

The first formula to enter into the cost breakdown summary sheet calculates the total of the actual costs for an item. To sum the actual costs (draws)

- Open the *Chapter 3 workbook*.
- Select cell **J9** in the Total Draws column. When you select a cell and enter a formula, the result of the formula will appear in the cell you have selected.
- Type the formula: =SUM(G9:I9).

To create this formula faster

- Select cell J9.
- Click the **AutoSum (Σ)** button on either the **Home** tab or the **Formula** tab (fig. 3.2).

Excel® enters the formula =SUM() into cell J9 and displays it in the formula bar. To edit formulas, click in the formula bar.

Excel® automatically selects cells F9:I9 as the sum range but you need to select G9:I9. To do this

- Click cell G9.
- Press and hold the left mouse button.
- Drag to cell I9.

Excel® automatically enters G9:I9 within the parenthesis. The formula should appear as follows:

$$=SUM(G9:I9)$$

Experiment with the AutoSum formula. You can AutoSum cells that are non-adjacent, or noncontiguous.

- Select the first cell or group of cells.
- While pressing the **Ctrl** key, select the next group or groups of cells.

Figure 3.3 shows how to AutoSum cells that are not in a single group.

Click the *Sum* worksheet tab in the Chapter 3 workbook. Notice that cell E1 already contains the AutoSum formula. To enter the AutoSum formulas into cell E2

- Select E2.
- Click the **AutoSum** button.
- Using the mouse, select cells A1 and A2.
- Press and hold **Ctrl**.
- Select B2, B3, C2, and D2.

Figure 3.2. AutoSum button

Figure 3.3. AutoSumming non-adjacent cells

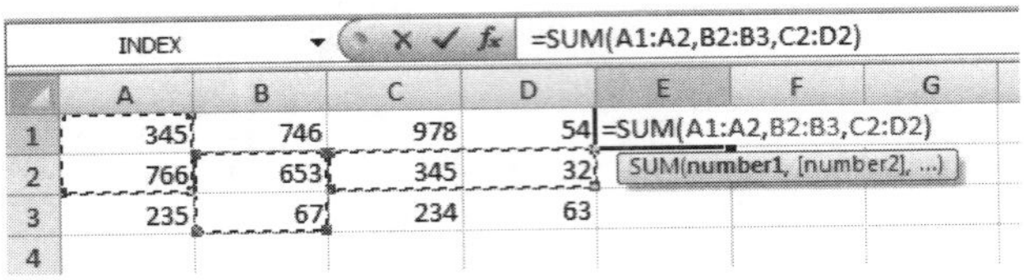

- Release the **Ctrl** key.
- Press **Enter**.

When you press **Enter**, a value of 2208 will appear in E2.

Now, go back to the *CostBreakdownSummary* worksheet and enter a formula to calculate the variance between the estimated costs and the total draws (the total actually paid).

- Select cell K9 by clicking on the cell.
- Type =F9−J9.

To perform the same operation even faster

- Select cell K9.
- Press the = key.
- Click F9.
- Press the **hyphen (-) key** (to the right of the zero or top right of the number key pad).
- Click J9.

Look at the formula bar as you click cells E8 and J8. Excel® enters them into the formula.

Copying Formulas

You have created the formulas to total the draws and calculate the variance between the estimated costs and the actual costs. You could repeat this procedure for rows 9–15, but there is an easier way. Simply copy the formulas as follows:

- Select cells **J9** and **K9**.
- **Move** the cursor to the fill handle in the bottom right corner of the selected cells (fig. 3.4). The cursor changes to a crosshair.

Figure 3.4. Copying formulas by selecting cells

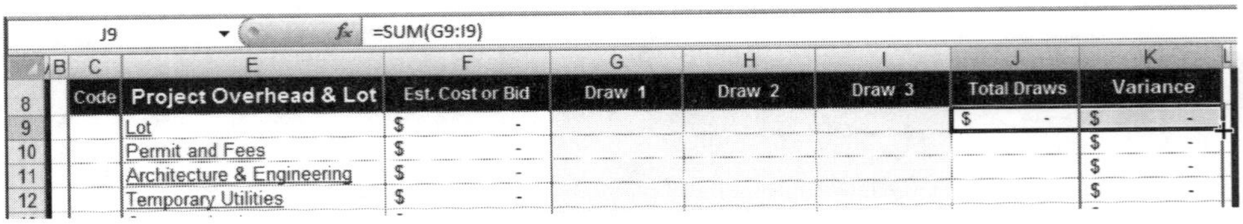

Figure 3.5. Copying formulas using fill

J9			f_x	=SUM(G9:I9)						
B	C	E	F	G	H	I	J	K	M
8	Code	**Project Overhead & Lot**	Est. Cost or Bid	Draw 1	Draw 2	Draw 3	Total Draws	Variance	
9		Lot	$ -				$ -	$ -	
10		Permit and Fees	$ 6,800.00	$ 4,200.00	$ 1,450.00	$ 950.00	$ 6,600.00	$ 200.00	
11		Architecture & Engineering	$ -				$ -	$ -	
12		Temporary Utilities	$ -				$ -	$ -	
13		Construction Loan	$ -				$ -	$ -	
14		Supervision	$ -				$ -	$ -	
15		Contingency	$ -				$ -	$ -	
16		Liability Insurance	$ -				$ -	$ -	
17		Real Estate Commission	$ -				$ -	$ -	
18									
19		**Subtotal - Project OH & Lot**							

- Click the **crosshair** and, while pressing the left mouse button, **drag** the fill handle to **K17** (fig. 3.5).

If you right click while dragging down with the fill handle, you get the options to "Fill Formatting Only" or "Fill Without Formatting."

In this example, the formula in cell J9 will be copied to cells J10–J17, and the formula in cell K9 will be copied to cells K10–K17. With the values entered into cells F10:I10 as shown, the total draws equal $6,600. The variance between the estimated cost of $6,800 and the total draws is $200.

To find the subtotal for all of the columns

- Select cells F19:K19.
- Click the **AutoSum** button.

Excel® automatically enters the summation formulas in cells F19:K19 (fig. 3.6).

The NOW Function

You may want to display the current date whenever the cost breakdown summary sheet is opened. You can quickly enter the **NOW** function into a formula by using the **Function Wizard** to the left of the formula bar on the Standard toolbar (fig. 3.7).

- Select cell J2.
- Click the **Function Wizard**. The **Insert Function** dialog box appears (fig. 3.8).

Figure 3.6. AutoSumming subtotals

F19			f_x	=SUM(F9:F18)						
B	C	E	F	G	H	I	J	K	
8	Code	**Project Overhead & Lot**	Est. Cost or Bid	Draw 1	Draw 2	Draw 3	Total Draws	Variance	
9		Lot	$ -				$ -	$ -	
10		Permit and Fees	$ 6,800.00	$ 4,200.00	$ 1,450.00	$ 950.00	$ 6,600.00	$ 200.00	
11		Architecture & Engineering	$ 2,500.00	$2,200.00	$ 350.00		$ 2,550.00	$ (50.00)	
12		Temporary Utilities	$ -				$ -	$ -	
13		Construction Loan	$ -				$ -	$ -	
14		Supervision	$ -				$ -	$ -	
15		Contingency	$ -				$ -	$ -	
16		Liability Insurance	$ -				$ -	$ -	
17		Real Estate Commission	$ -				$ -	$ -	
18									
19		**Subtotal - Project OH & Lot**	$ 9,300.00	$ 6,400.00	$ 1,800.00	$ 950.00	$ 9,150.00	$ 150.00	

Figure 3.7. Function Wizard

Figure 3.8. Insert Function dialog box

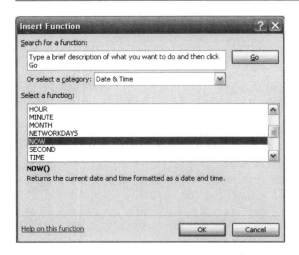

Figure 3.9. NOW function

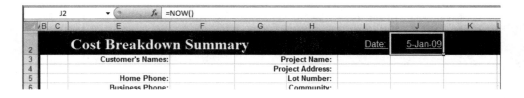

- Select **Date & Time.**
- Select **NOW.**
- Click **OK** or double click **NOW.**

Excel® inserts the formula **=NOW()** into cell J2 (fig. 3.9).

Formatting the Date

If J2 has not been formatted, you will need to format the cell to display the date as you wish to see it.

- Right click J2.
- Click **Format Cells**.
- Choose the **Number** tab.
- Under Category, select **Date**.
- Choose the desired format or create a custom format (fig. 3.10).

Excel® displays the chosen current date and/or time format in cell J2 (fig. 3.11).

If you press **Ctrl + ;** (the control key and the semicolon simultaneously), Excel® automatically enters the current date into the active cell. If you press **Ctrl + Shift + ;** (the control key, the shift key, and the semicolon) Excel® enters the current time.

Figure 3.10. Customizing Date Format

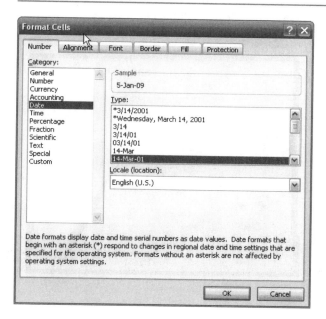

Figure 3.11. Spreadsheet with data displayed

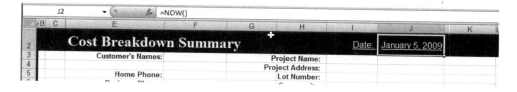

Formatting for Accounting

You can format cells to display numbers as currency—with the dollar sign, decimal point, and commas to indicate thousands. However, if you want each of these three symbols to align, you must format the cells to Accounting Number style:

- Select the cells that you want to change.
- Click the **Accounting Number** format button (**$**) on the **Home** tab of the ribbon.

Increasing or Decreasing the Number Decimal Places

You can also change the number of places beyond the decimal:

- Select the cells that you want to change.
- Click the **Increase Decimal** or **Decrease Decimal** buttons (fig. 3.12).

    You should create and copy your formulas to the desired locations before formatting your spreadsheet because when you use **AutoFill** to copy formulas, it will copy the formatting too. If you set up the formatting first, you may have to redo many of the formats.

Figure 3.12. Accounting number format and Increase or Decrease Decimal buttons

---

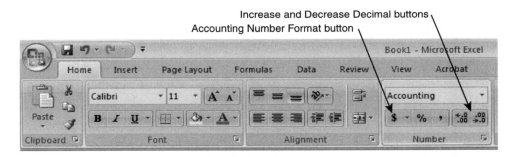

Paste Special

Instead of right clicking to access the pop-up menu on the fill handle, you can use the **Paste Special** command to copy formulas (but not formats).

- Select the cell or cells you want to copy.
- Click the **Copy** button on the **Home** tab or right click the cell or cells and click copy on the menu.
- Select the cell or range of cells to which you want to copy the formula.
- Click the **down arrow** on the **Paste** command.
- Choose **Paste Special** from the drop-down menu (fig. 3.13).
- Choose **Formulas** in the pop-up dialog box (fig. 3.14).
- Click **OK** or press **Enter**.

Excel® copies the formulas to the new cell locations without copying the formats.

Figure 3.13. Paste menu

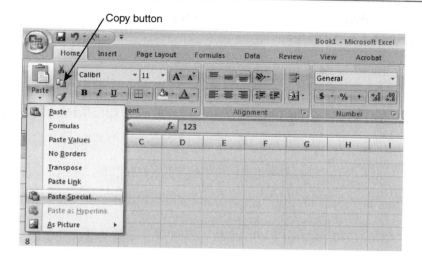

Figure 3.14. Paste Special dialog box

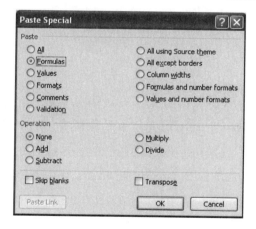

Paste Special Operators

The operation functions save time in performing calculations, such as price increases. You can find these functions in the **Paste Special** dialog box.

Here's how to use them: Your roofing supplier notifies you that all roofing material prices will increase by 8% for the coming year. Recalculating figures in situations like this used to be a nightmare. But you can update the costs quickly using the **Multiply** operator.

Figure 3.15 shows unit costs located in column C. To quickly add 8% to all of the current costs

Figure 3.15. Paste Special: Multiply operation

	A	B	C	D
1	Roofing Material Database			
2				
3	Description	Unit	Unit Cost	Percentage to increase
4	Asphalt Shingles - 25 yr. 3 Tab	SQ	$ 36.00	1.08
5	Asphalt Shingles - 30 yr. 3 Tab	SQ	$ 39.50	
6	Asphalt Shingles - 30 yr. Architectural	SQ	$ 54.00	
7	Cedar Shakes #1 medium handsplits	SQ	$ 132.00	
8	Cedar Shingles #1	SQ	$ 162.00	
9	Metal Roofing	SQ	$ 108.00	
10	Eagle Tile	SQ	$ 126.00	
11	--			
12	Deliver & Stock shingles	SQ	$ 4.20	
13	Drip Edge-metal	Ea	$ 4.68	
14	15# Tar Paper (Asphalt-impregnated Felt)	4-Sq Roll	$ 13.14	
15	30# Tar Paper (Asphalt-impregnated Felt)	2-Sq Roll	$ 13.14	
16	Ice and Water Shield - 3' x 75'	2-Sq Roll	$ 95.40	
17	Starter Strip	SQ	$ 39.50	

- Enter 1.08 into any cell (D4 in this case).
- **Copy** the cell value.
- Select cells C3:C? (Replace the "?" with the row number of the last cell you want to include in the price update—the cell at the end of the list of items.)
- Click the **Paste Special** command.
- Select the **Multiply** operation.
  - Click **OK** or press **Enter**.

Excel® multiplies each value in column C by the value in D4 and returns the updated prices to column C (fig. 3.16).

Figure 3.16. Updated costs

	A	B	C	D
1	Roofing Material Database			
2				
3	Description	Unit	Unit Cost	Percentage to increase
4	Asphalt Shingles - 25 yr. 3 Tab	SQ	$ 38.88	1.08
5	Asphalt Shingles - 30 yr. 3 Tab	SQ	$ 42.66	
6	Asphalt Shingles - 30 yr. Architectural	SQ	$ 58.32	
7	Cedar Shakes #1 medium handsplits	SQ	$ 142.56	
8	Cedar Shingles #1	SQ	$ 174.96	
9	Metal Roofing	SQ	$ 116.64	
10	Eagle Tile	SQ	$ 136.08	
11	--			
12	Deliver & Stock shingles	SQ	$ 4.54	
13	Drip Edge-metal	Ea	$ 5.05	
14	15# Tar Paper (Asphalt-impregnated Felt)	4-Sq Roll	$ 14.19	
15	30# Tar Paper (Asphalt-impregnated Felt)	2-Sq Roll	$ 14.19	
16	Ice and Water Shield - 3' x 75'	2-Sq Roll	$ 103.03	
17	Starter Strip	SQ	$ 42.66	

Paste Special/Transpose

The **Transpose** function, located at the bottom of the Paste Special dialog box, helps you quickly transform lists from vertical columns to horizontal rows or vice versa. In figure 3.17, for example, you may want to arrange the vertically placed city names (on the *Transpose* worksheet tab) horizontally. To transpose data, follow these steps:

- **Copy** the cells that contain the list of names.
- **Right** click the starting cell of the horizontal segment (A1 in this case) and Excel® displays a menu.
- Click **Paste Special.**
- Check the box next to **Transpose.**
- Click **OK.**

Figure 3.17. Paste Special: Transpose

	A	B	C	D	E	F	G	H	I	J	K
1	Atlanta	Dallas	Seattle	St. Louis	Richmond	San Diego	Buffalo	Orlando	St. Paul	Boise	Tulsa
2											
3	Atlanta										
4	Dallas										
5	Seattle										
6	St. Louis										
7	Richmond										
8	San Diego										
9	Buffalo										
10	Orlando										
11	St. Paul										
12	Boise										
13	Tulsa										

Sorting Data

You can sort the list of cities in figure 3.18 in alphabetical order by state by using the **Sort** function.

- Select the list of cities and states (include the headings).
- On the **Data** tab, click **Sort.**

The Sort dialog box appears. You are asked to choose how you would like to sort—by city (column A) or state (column B). Notice the two buttons to the left of the **Sort** button. One will sort in ascending order (A to Z); the other will sort in descending order (Z to A). In cases like the example, with multiple cities listed for each state, you may wish to sort data according to secondary criteria.

- Click **Add Level** at the upper left corner of the Sort dialog box to add secondary or tertiary sorting criteria. Excel® first sorts the states alphabetically and then sorts the cities alphabetically within their states (fig. 3.19).

Figure 3.18. Sorting data

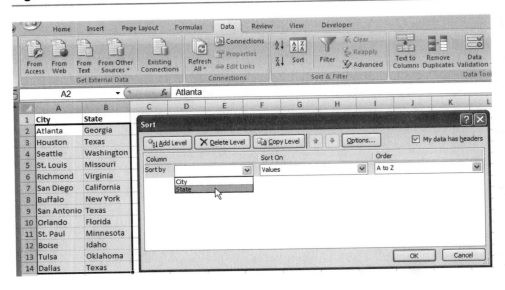

Figure 3.19. Data after sorting

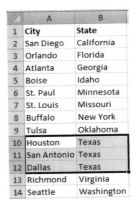

Make a Cost Breakdown Summary Sheet

This chapter discussed how to create a cost breakdown summary sheet, how to write formulas, and how to format worksheets. If you haven't already done so, take time now to complete your own cost breakdown summary sheet using the instructional workbook for Chapter 3 as a guide.

Eliminating Steps by Using Detail Sheets

4

Detail sheets (or work-up sheets) are the workhorses of spreadsheet estimating. They contain the take-off quantities as well as formulas and functions to automate most of the estimating process. The formulas and functions you incorporate into detail sheets calculate totals so you don't have to. A well-designed detail sheet can eliminate most of the work of estimating and significantly reduce estimating time.

Detail sheets generally will look like the example shown in figure 4.1. The estimator enters a description for each item, the item quantity by a specific unit of measure, and the unit costs. Excel® multiplies the unit costs by the quantities and adds the totals to the detail sheets.

Databases support detail sheets. You can write formulas to help Excel® locate specific information within these databases and add it to your detail sheets.

There are two Chapter 4 workbooks on the CD: Chapter 4-Book 1 and Chapter 4-Book 2. Use Chapter 4-Book 1 to work through the examples in this chapter. Use Chapter 4-Book 2 (a completed workbook) to check your work as you create your own detail and database sheets. Book 2 includes worksheets that show intermediate steps toward the final product.

In this chapter, you will create a roofing detail sheet and two roofing databases—one database for material costs and one for labor costs. The two databases, *RoofingMatDB* and *RoofingLaborDB*, could have been included in one worksheet, but they have been separated in order to clearly demonstrate each concept in this chapter. You will also use the *CostBreakdownSummary* and *KeyInfo* worksheets in Book 1.

 Use a naming convention that will help you organize and keep track of your worksheets. For example, you can name detail sheets by construction phase or activity, and then add "DB" to the end of the name for the worksheet that contains the corresponding database.

Figure 4.1. Detail sheet

	A	B	C	D	E
1	Roofing				
2	Description	QTY	Unit	$/Unit	Sub-Total $
3	Asphalt Shingles - 30 yr. 3 Tab		Sq	$ 38.00	$ -
4	Deliver & Stock shingles		Sq	$ 3.50	$ -
5	Drip Edge-metal		Ea	$ 3.90	$ -
6	15# Tar Paper (Asphalt-impregnated Felt)		4-Sq Roll	$ 10.95	$ -
7	Starter Strip		Sq (T)	$ 38.00	$ -
8	Hip & Ridge Cap		Sq (T)	$ 38.00	$ -
9	Ridge Vents		Lf	$ 2.30	$ -
10	Step Flashing 3 x 4 x 7 (22' / Bndl.)		Bndl of 50	$ 12.95	$ -
11	L Metal 4" x 4" x 10'		Ea	$ 4.79	$ -
12	Plastic Caps 1/2 lb./sq		Pail	$ 34.95	$ -
13	Coil Roofing Nails 1" - 1-1/4"		Ea	$ 37.95	$ -
14	Labor		SQ	$ 38.00	$ -
15					
16					
17				Total Cost:	$ -

Extending Item Costs

To calculate the item total costs (column E), you must create a formula to multiply the product by the unit price as follows:

- Click E3.
- Type **=B3*D3** (To avoid typing errors, after entering the = sign in cell E3, click cell B3 to automatically enter B3 into the formula. Do the same for D3 after entering the * sign.)
- **Copy** the formula to cells E4:E15 by dragging E3's fill handle to and including cell E15.

To calculate the total of all the roofing items

- Select cell E17.
- Click **AutoSum**.

Enter values into the QTY column to check your formulas. Your practice roofing detail sheet should now match the Roofing detail sheet shown in figure 4.2.

Figure 4.2. Detail sheet with price extensions

	A	B	C	D	E
1	Roofing				
2	Description	QTY	Unit	$/Unit	Sub-Total $
3	Asphalt Shingles - 30 yr. 3 Tab	29	Sq	$ 38.00	$ 1,102.00
4	Deliver & Stock shingles	32	Sq	$ 3.50	$ 112.00
5	Drip Edge-metal	24	Ea	$ 3.90	$ 93.60
6	15# Tar Paper (Asphalt-impregnated Felt)	8	4-Sq Roll	$ 10.95	$ 87.60
7	Starter Strip	1	Sq (T)	$ 38.00	$ 38.00
8	Hip & Ridge Cap	2	Sq (T)	$ 38.00	$ 76.00
9	Ridge Vents	80	Lf	$ 2.30	$ 184.00
10	Step Flashing 3 x 4 x 7 (22' / Bndl.)	2	Bndl of 50	$ 12.95	$ 25.90
11	L Metal 4" x 4" x 10'	3	Ea	$ 4.79	$ 14.37
12	Plastic Caps 1/2 lb./sq	2	Pail	$ 34.95	$ 69.90
13	Coil Roofing Nails 1" - 1-1/4"	2	Ea	$ 37.95	$ 75.90
14	Labor	32	SQ	$ 38.00	$ 1,216.00
15					
16					
17				Total Cost:	$ 3,212.72

Automating the Detail Sheet

Adding databases to your workbooks—lists that contain important information such as item descriptions, units of measure, and costs per unit—can further automate your detail sheets and speed up and improve estimating. A few preliminary steps will help you automate the detail sheet.

First, create a database with named ranges to use in the formulas and functions that will be part of the detail sheet. The quantity column in a database can store either default quantities or formulas to automatically calculate quantities. These formulas are one of the most powerful aspects of your estimating spreadsheet. Figure 4.3 shows a simplified roofing material database.

In Chapter 1, you learned how to name ranges of cells. Let's apply that knowledge to name the roofing database.

Figure 4.3. Roofing database

	A	B	C	D
1	**Roofing Material Database**			
2	**Description**	QTY	Unit	$/ Unit
3	Asphalt Shingles - 25 yr. 3 Tab		SQ (T)	$ 30.00
4	Asphalt Shingles - 30 yr. 3 Tab		SQ (T)	$ 38.00
5	Asphalt Shingles - 30 yr. Architectural		SQ (T)	$ 48.43
6	Cedar Shakes #1 medium handsplits		SQ (T)	$ 110.00
7	Cedar Shingles #1		SQ (T)	$ 135.00
8	Metal Roofing		SQ (T)	$ 90.00
9	Eagle Tile		SQ (T)	$ 105.00
10	--------------			
11	Deliver & Stock shingles		SQ	$ 3.50
12	Drip Edge-metal		Ea	$ 3.90
13	15# Tar Paper (Asphalt-impregnated Felt)		4-Sq Roll	$ 10.95
14	30# Tar Paper (Asphalt-impregnated Felt)		2-Sq Roll	$ 10.95
15	Ice and Water Shield - 3' x 75'		2-Sq Roll	$ 77.00
16	Starter Strip		SQ (T)	$ 38.00
17	Hip Cap		SQ (T)	$ 38.00
18	Ridge Cap		SQ (T)	$ 38.00
19	Architectural Hip Cap		Box	$ 30.00
20	Architectural Ridge Cap		Box	$ 30.00
21	Ridge Vents		Lf	$ 2.30
22	Turtle-back Vents 10 x 14 (1 SF net area)		Ea	$ 15.00
23	Step Flashing 3 x 4 x 7 (22' / Bndl.)		Bndl of 50	$ 12.95
24	18" x 10' Valley Flashing		Ea	$ 5.50
25	L Metal 4" x 4" x 10'		Ea	$ 4.79
26	Plastic Caps 1/2 lb./sq		Pail	$ 34.95
27	Simplex Nails-1/2 lb./sq		Lb	$ 0.85
28	Roofing Nails - 1-1/4" - 2 lb./sq		Lb	$ 0.62
29	Coil Roofing Nails 1" - 1-1/4"		Pail	$ 37.95
30	8d galvanized box nails		Lb	$ 0.95
31	6d galvanized box nails		Lb	$ 0.95
32	4d galvanized box nails		Lb	$ 0.95
33				
34	Labor		SQ	$ 38.00
35				

Figure 4.4. Naming the roofing database

RoofingDB	▼	f_x	Asphalt Shingles - 25 yr. 3 Tab	
	A	B	C	D
1	Roofing Material Database			
2	Description	QTY	Unit	$/ Unit
3	Asphalt Shingles - 25 yr. 3 Tab		Sq	$ 30.00
4	Asphalt Shingles - 30 yr. 3 Tab		Sq	$ 38.00
5	Asphalt Shingles - 30 yr. Architectural		Sq	$ 48.43
6	Cedar Shakes #1 medium handsplits		Sq	$ 110.00
7	Cedar Shingles #1		Sq	$ 135.00
8	Metal Roofing		Sq	$ 90.00
9	Eagle Tile		Sq	$ 105.00
10	-----------------------------------			
11	Deliver & Stock shingles		Sq	$ 3.50
12	Drip Edge-metal		Ea	$ 3.90
13	15# Tar Paper (Asphalt-impregnated Felt)		4-Sq Roll	$ 10.95
14	30# Tar Paper (Asphalt-impregnated Felt)		2-Sq Roll	$ 10.95
15	Ice and Water Shield - 3' x 75'		2-Sq Roll	$ 77.00

- Click the *RoofingMatDB* worksheet tab in Chapter 4-Book 1.
- Select cells A3:D35. (Leaving a blank line at the bottom of a database will allow users to insert additional items at the bottom of the table.)
- In the **Name** box (to the left of the formula bar), enter *RoofingDB* (fig. 4.4).
- Press **Enter**.

Name the range of cells that includes the list of descriptions for the roofing items as follows:

- Select cells A3:A35.
- Type *RoofingList* in the **Name** box.
- Press **Enter**.

Notice that the roofing database has four columns:

1. Column A holds the description data.
2. Column B stores the default quantities or the formulas that can be used to calculate the quantities automatically.
3. Column C holds the units of measure for each of the roofing items.
4. Column D stores the unit costs.

Formulas that look up items from a database will require both the column references in the cell address (A, B, C ...) and their numerical order in your worksheets (1, 2, 3 ...). The formulas refer to the numbers as column index numbers (Col_index_number).

Data Validation: Inputting Data

Inputting data is the most time-consuming part of estimating. Fortunately, Excel® allows you to select items from a list instead of having to type them in.

By naming cell ranges, you have already formatted the roofing database worksheet to enable **Data Validation**. Recall that when you created the roofing database, you named the cell range A3:D35 *RoofingDB* and the range A3:A35 *RoofingList*. You will use *RoofingList* with Data Validation to input data using a drop-down list.

- In the *Roofing* detail sheet, select A3.
- Click the **Data** tab.
- Click **Data Validation** in the **Data Tools** section (fig. 4.5).

Figure 4.5. Accessing Data Validation from the Data tab

Validation is a method to control what information goes into a cell. There are several validation criteria. You want to specify information that is on a list.

- Under **Allow:** choose **List**.
- In the box under **Source:** you can manually enter the range of cells where the list is located or, because you previously named the range of cells, you can simply type **=RoofingList** in the box under **Source** (fig. 4.6). An even faster option is to press the F3 key and select *RoofingList* from the list of named cell ranges.

Figure 4.6. Data Validation: Allow List

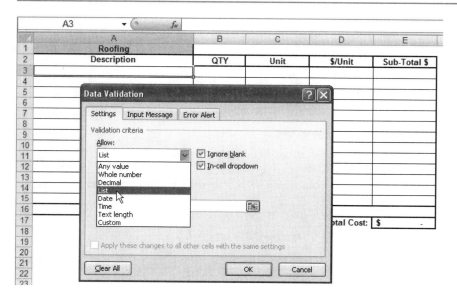

Figure 4.7. Data Validation source

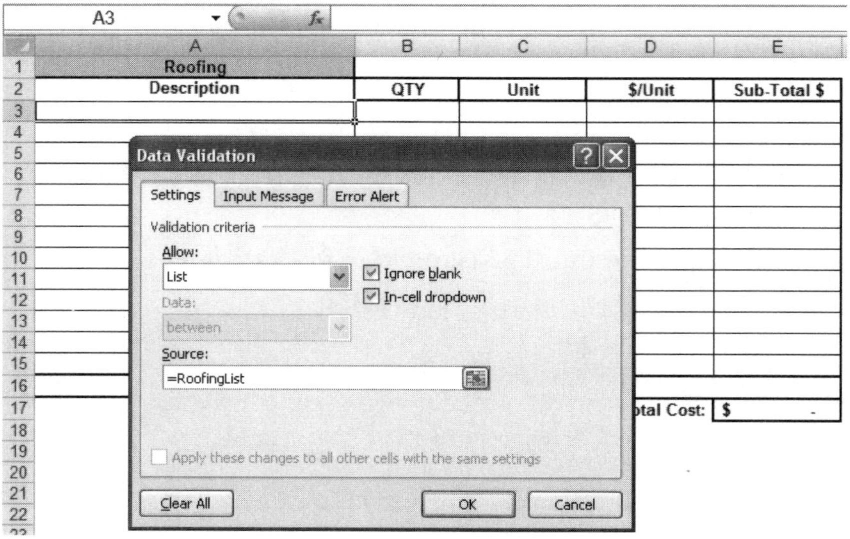

- Click **OK**. After you click off the cell and select it again, you will see a small downward-facing arrow next to cell A3.
- Click the arrow. A drop-down list of roofing descriptions will appear (fig. 4.7).
- Select **Asphalt Shingles–30 yr. 3 Tab** from the list.

Excel® enters the shingle description automatically into cell A3.

- Using the fill handle to drag, **copy** cell A3 down to cells A4:A15.
- Choose new item descriptions from the drop-down list for cells A4:A15 (fig. 4.8).

Figure 4.8. Data Validation: Drop-down list

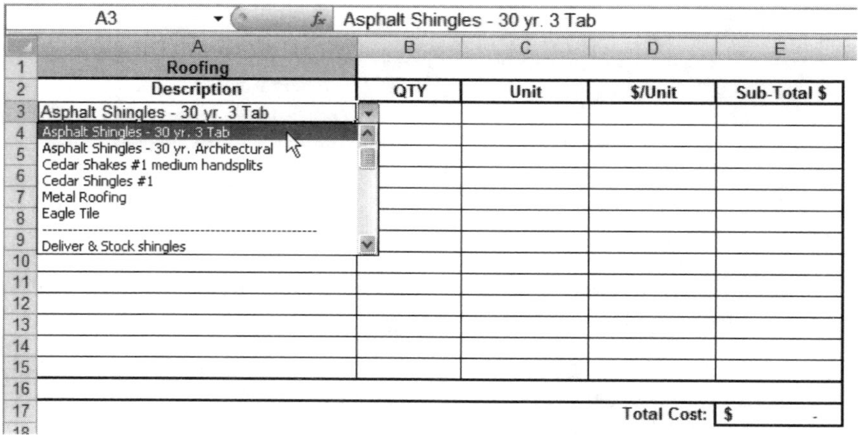

Vertical Lookup

You can further automate data entry by using the Vertical Lookup (**VLOOKUP**) function to locate and retrieve data from another worksheet or workbook. To insert the VLOOKUP function:

- Select D3 (fig. 4.9).
- Click the **Function Wizard.**
- Under **Select a category,** choose **Lookup & Reference.**
- Under **Select a function,** choose **VLOOKUP.** 🖉 You can type "v" under **Search for a function** to be directed to the words that start with the letter V.
- Click **OK** or press **Enter.**

Figure 4.9. Data Validation completed

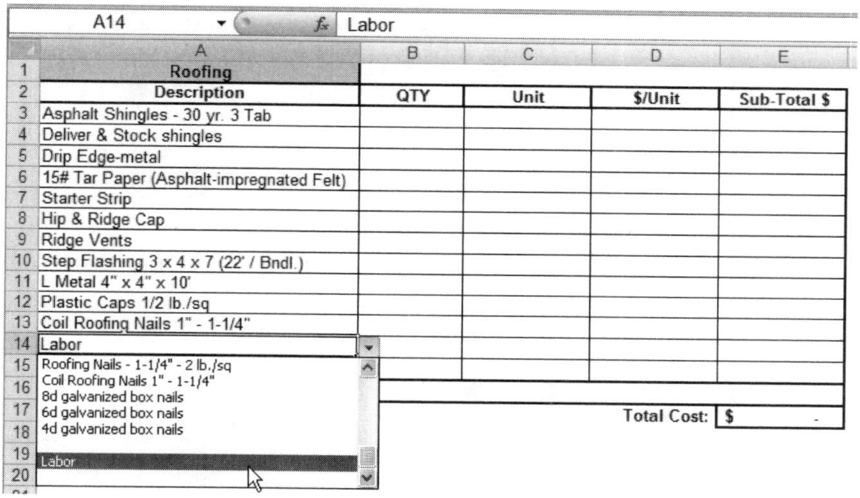

The VLOOKUP dialog box pops up (fig. 4.10) and prompts you to enter the Lookup_value, Table_array, Col_index_number, and the Range_lookup (fig. 4.11).

Lookup_value is the value that VLOOKUP searches for in the first column of a specified database. In other words, the value VLOOKUP returns depends on the Lookup_value. VLOOKUP matches whatever value is in the Lookup_value to a value in the first column of the roofing database. If VLOOKUP doesn't find an exact match, then VLOOKUP returns an error value **#N/A.**

- Define the Lookup_value as A3. Click in the box next to Lookup_value in the dialog box and enter A3 (or click cell A3 to enter A3 automatically into the Lookup_value box).

Table_array tells VLOOKUP which database or table array to use. To find the cost per unit ($/Unit) for roofing, you want VLOOKUP to search the roofing database (*RoofingDB*).

Figure 4.10. Initiating VLOOKUP

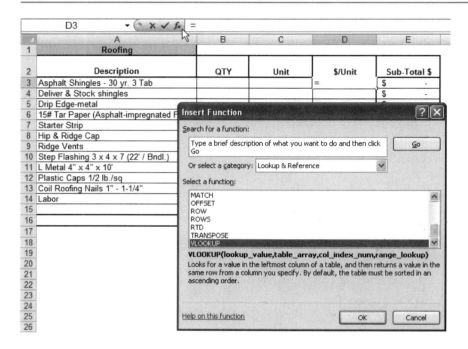

Figure 4.11. VLOOKUP arguments

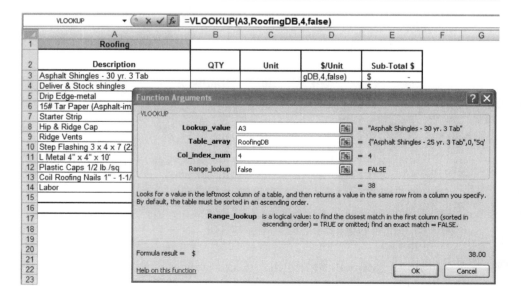

- Either type *RoofingDB* in the box next to Table_array, or press **F3** to display the list of named cells in the workbook.

- After the list is displayed, click *RoofingDB* and **OK** or **Enter**, or simply double click *RoofingDB*.

Col_index_number is the column number in the database that holds the information you want VLOOKUP to retrieve. You want VLOOKUP to find the unit cost for a specific item in the roofing database. There are four columns in the named range *RoofingDB*. The first column stores item descriptions, the second column holds quantity defaults or formulas, the third column holds the units, and the fourth column stores the unit costs ($/Unit).

- Enter **4** in the Col_index_number box.

Range_lookup tells VLOOKUP whether to find an exact match between the Lookup_value (the value in cell A3) and the descriptions in column 1 of the database or to pick the description in column 1 that is most similar to the Lookup_value.

If you type **false** in the Range_lookup box, VLOOKUP seeks an exact match with the Lookup_value from the choices in column 1 of the database. If you type **true**, VLOOKUP selects the nearest value to the Lookup_value from the choices in column 1 of the database.

- Type **false** in the Range_Lookup box. (In estimating, you want an exact match so your cost per unit is accurate.)

You can use the Function Wizard to help you create formulas.

- Select D3 and notice that the **Function Wizard** created the following formula, which is displayed in the formula bar:

=VLOOKUP(A3,RoofingDB,4,FALSE)

- Click any of the boxes of the **Function Wizard**. Excel® prompts you with instructions (near the bottom of the dialog box) to help you correctly enter the information into the box. If you don't understand how to create a formula, click **Help on this function** located in the bottom left corner of the **Function Wizard** dialog box.

- After entering the information into the **Function Wizard**, click **OK**.

With "Asphalt Shingles—30 yr. 3 Tab" correctly entered as the description, Excel® displays the unit cost of $38.00 in cell D4 (or simply 38 if you have not formatted the cell to display numbers in currency style). If you were to enter an incorrect description, such as "Asphalt Shingles—31 yr. 3 Tab," Excel® would display the #N/A error value.

After you have tested the VLOOKUP formula, copy it to D4:D15. Remember, if you right click the fill handle to copy the formula, the pop-up menu offers the option to **Fill Without Formatting**, which maintains the original cell formatting.

If you don't enter a value in the Description column, VLOOKUP returns an error value (fig. 4.12), which causes errors in both the Subtotal column and the Total Cost cell. An #N/A error in the $/Unit column is not a problem because it indicates that the Description column contains an incorrect value or no value.

Figure 4.12. VLOOKUP results

	D3	▾	f_x	=VLOOKUP(A3,RoofingDB,4,FALSE)	
	A	B	C	D	E
1	Roofing				
2	Description	QTY	Unit	$/Unit	Sub-Total $
3	Asphalt Shingles - 30 yr. 3 Tab	29		$ 38.00	$ 1,102.00
4	Deliver & Stock shingles			$ 3.50	$ -
5	Drip Edge-metal			$ 3.90	$ -
6	15# Tar Paper (Asphalt-impregnated Felt)			$ 10.95	$ -
7	Starter Strip			$ 38.00	$ -
8	Hip & Ridge Cap			$ 38.00	$ -
9	Ridge Vents			$ 2.30	$ -
10	Step Flashing 3 x 4 x 7 (22' / Bndl.)			$ 12.95	$ -
11	L Metal 4" x 4" x 10'			$ 4.79	$ -
12	Plastic Caps 1/2 lb./sq			$ 34.95	$ -
13	Coil Roofing Nails 1" - 1-1/4"			$ 37.95	$ -
14	Labor			$ 38.00	$ -
15				#N/A	#N/A
16					
17				Total Cost:	#N/A
18					

However, you don't want an error value in the Subtotal column because that will cause an error in the Total Cost as well.

One solution is to enter an **IF** function in the Subtotal column (fig. 4.13). In other words, you tell Excel®, "If A3 is blank, enter a zero into the Subtotal column. If not, calculate the product of the quantity (B3) and the cost per unit (D3)." The IF function is a conditional statement that takes the following form:

=IF(logical_test, [value_if_true], [value_if_false])

- Enter the following formula into cell E3:

=IF(ISBLANK(A3),0,B3*D3).

Figure 4.13. Clearing error results with IF function

	E3	▾	f_x	=IF(ISBLANK(A3),0,B3*D3)				
	A	B	C	D	E	F	G	H
1	Roofing							
2	Description	QTY	Unit	$/Unit	Sub-Total $			
3	Asphalt Shingles - 30 yr. 3 Tab	29		$ 38.00	$ 1,102.00			
4	Deliver & Stock shingles			$ 3.50	$ -			
5	Drip Edge-metal			$ 3.90	$ -			
6	15# Tar Paper (Asphalt-impregnated Felt)			$ 10.95	$ -			
7	Starter Strip			$ 38.00	$ -			
8	Hip & Ridge Cap			$ 38.00	$ -			
9	Ridge Vents			$ 2.30	$ -			
10	Step Flashing 3 x 4 x 7 (22' / Bndl.)			$ 12.95	$ -			
11	L Metal 4" x 4" x 10'			$ 4.79	$ -			
12	Plastic Caps 1/2 lb./sq			$ 34.95	$ -			
13	Coil Roofing Nails 1" - 1-1/4"			$ 37.95	$ -			
14	Labor			$ 38.00	$ -			
15				#N/A	$ -			
16								
17				Total Cost:	$ 1,102.00	Copy Cells		
18						Fill Series		
19						Fill Formatting Only		
20						Fill Without Formatting		
21						Fill Days		
22								

Figure 4.14. Finding units with VLOOKUP

	A	B	C	D	E
	C3		▼ *fx* =VLOOKUP(A3,RoofingDB,3,FALSE)		
1	Roofing				
2	Description	QTY	Unit	$/Unit	Sub-Total $
3	Asphalt Shingles - 30 yr. 3 Tab	29	Sq	$ 38.00	$ 1,102.00
4	Deliver & Stock shingles		Sq	$ 3.50	$ -
5	Drip Edge-metal		Ea	$ 3.90	$ -
6	15# Tar Paper (Asphalt-impregnated Felt)		4-Sq Roll	$ 10.95	$ -
7	Starter Strip		Sq (T)	$ 38.00	$ -
8	Hip & Ridge Cap		Sq (T)	$ 38.00	$ -
9	Ridge Vents		Lf	$ 2.30	$ -
10	Step Flashing 3 x 4 x 7 (22' / Bndl.)		Bndl of 50	$ 12.95	$ -
11	L Metal 4" x 4" x 10'		Ea	$ 4.79	$ -
12	Plastic Caps 1/2 lb./sq		Pail	$ 34.95	$ -
13	Coil Roofing Nails 1" - 1-1/4"		Ea	$ 37.95	$ -
14	Labor		Sq	$ 38.00	$ -
15			#N/A	#N/A	$ -
16					
17				Total Cost:	$ 1,102.00

The formula states, "If A3 is blank, then enter the value 0 into cell E3. Otherwise, enter the product of B3 and D3 into cell E3."

- **Copy** the formula through cell E15.

You also can enter a VLOOKUP formula in Column C to look up the units of the items in the database as follows (fig. 4.14):

- Select C3.
- Enter the formula **=VLOOKUP(A3,RoofingDB,3,FALSE)**.

Again, use the Function Wizard to help create the formula. The only difference between the formula in cells C3 and D3 is that the Col_index_numbers are, respectively, 3 and 4.

- Test the formula and then **copy** it to cells C4:C15.

Calculating Quantities

You will save time by programming Excel® to calculate quantities automatically based on a few values you input from the take-off. Then, even when you are building from a variety of plans, you will only need to input the take-off quantities. Excel® automatically calculates materials and labor. You will no longer have to use a calculator to compute the order quantities for the items on the estimate.

Streamlining the Process

Estimating the roofing for a building can be time consuming. Many builders and remodelers describe their estimating process as follows: "I pull out the elevation plans, measure the area of each slope of the roof, and add them together," or "I climb on the roof and within 30 minutes or so, I can get the exact measurements of the roof." The problem is that the estimator will spend half an hour driving to the job and getting on the roof, another half hour driving back to the office, and another half hour calculating quantities and making a materials list. That's two hours that could be used to create the estimate!

Seasoned framers can cut rafters for an entire hip roof without getting on the roof. They calculate ridge cuts and bird's mouths and cut the rafters on the ground. The ridge is then lifted into place, the rafters are installed, and the roof is trimmed out. I have also seen many framers set the ridge and then climb up and down the ladder several times to measure, mark, cut, and check numerous times before they discover a rafter that they can use for a pattern. The process is even worse for hips, jacks, and valleys.

Estimating Roofing

Fortunately, the same principles that allow the seasoned framers to calculate rafters on the ground can be used to calculate the squares of shingles needed for the roof. Construction professionals are amazed when they see how rapidly and accurately they can estimate roofing using Excel®.

Flat vs. Sloped

Figure 4.15 is a plan view of a typical hip roof. The roof dimensions in the figure are to the fascia. From this view, the roof appears flat. To calculate the number of shingles needed for this flat roof

Figure 4.15. Plan view of roof

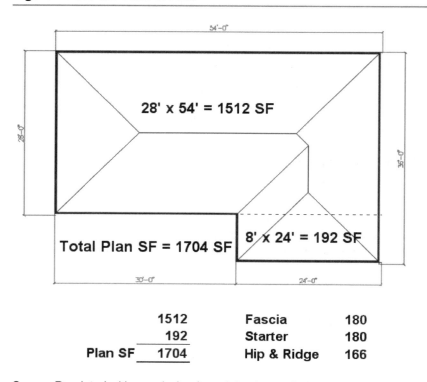

		Fascia	180
	1512	Starter	180
	192		
Plan SF	1704	Hip & Ridge	166

Source: Reprinted with permission from John Jones, Softplan

1. Determine the roof area.
2. Divide this number by 100.
3. Add the number of starter shingles to the result. (There are no hip and ridge shingles on a flat roof.)

For ease of calculation, the plan-view area of the roof is divided into two rectangles. One is 28′ × 54′ and the other is 8′ × 24′. The total plan view or flat SF is 1,704 SF (1,512 SF + 192 SF). If the roof were flat, you would need approximately 17⅓ squares of shingles—not including starter strip or waste—to cover it. Sloped roofs require more shingles than flat roofs for the same footprint area. As the slope increases, so does the required number of shingles.

This roof has a 6/12 slope. That is, for every 12″ of run (horizontal distance), the roof rises 6″. If you know the rise and the run of a roof slope (fig. 4.16), you can determine the length of the sloped section using the *Pythagorean theorem*. You will use the length of the sloped section to create a factor to determine the quantity of shingles you need to cover each flat SF of roof, taking the degree of slope into account.

The *Pythagorean theorem* states that the square of the length of the hypotenuse of a right triangle (the side opposite the right angle) equals the sum of the squared lengths of the other two sides, or

$$a^2 + b^2 = c^2$$

The equation for the roof pictured is as follows: The squared area of side **a** (12″ × 12″ = 144″²) plus the squared area of side **b** (6″ × 6″ = 36″²) is equal to the squared area of **c** (180″²).

Use the square root of the area of the square (180″ in this case) to find the length of side **c** in inches.

$$c = \sqrt{a^2 + b^2} = \sqrt{6″^2 + 12″^2} = \sqrt{36″ + 144″} = \sqrt{180″} = 13.42″ = 1.118′$$

The slope factor, 13.42″, equals 1.118′ (13.42″ / 12″ = 1.118′). For a 6/12 roof slope, this equation says that for every 1′ of run (horizontal distance) the length

Figure 4.16. Applying the Pythagorean theorem

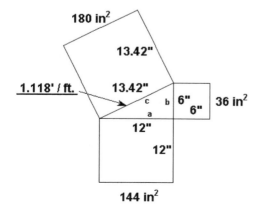

along the slope of the roof is 13.42 inches. For every SF measured horizontally (the plan view of the roof), the area of the roof is 1.118'. This slope factor (1.118) can be used to convert the flat SF of an area to sloped SF area. The roof example has 1,704 of flat SF. To calculate the sloped roof area, multiply the flat SF by the slope factor:

$$1,704 \text{ SF} \times 1.118 = 1,905 \text{ SF}$$

Waste Factors

Not counting starter strip, hip and ridge shingles, or waste, you would need to order 19⅓ squares of shingles (3 bundles per square) to cover this 6/12 sloped roof. You can use the same formula to calculate roof sheathing. For roof sheathing, add a 20% waste factor to allow for cuts and overbuilds. (For gable roofs, use a waste factor of 2%–5% and for hip roofs, 10%.)

You should name the cells that hold the quantity take-off inputs. (For example, name cell B3 PlanSF, cell B4, Slope, etc.). You will use these names in the quantity-calculation formulas. Figure 4.17 shows a formula to calculate the slope factor for any slope. The notation SQRT will calculate the square root of whatever follows it in parentheses. Cell B5 has been named RoofSlope. The ^ symbol is the power sign, so the RoofSlope is being raised to the power of 2. In other words, it is being squared, which means RoofSlope × RoofSlope.

Figure 4.17. Slope factor formula

RoofSlopeFactor	*fx*	=SQRT(RoofSlope^2+144)/12	
	A	B	C
1	**Roofing Detail Sheet**		
2	Take-off Inputs		
3	Plan SF of Roof	1704	
4	Slope (enter the rise only)	6	
5	Fascia LF	180	
6	Starter LF	180	
7	Ridge LF	34	
8	Hip LF (A)	68	
9			
10			
11	Calculated		
12	Slope Factor	1.118033989	
13	Squares		

After you calculate the slope factor, you can easily determine the necessary quantity of squares as follows:

1. Multiply the SF of the roof on the home plan by the slope factor.
2. Divide the result by 100 to change the result to squares.

The basic formula is

$$=\text{RoofingPlanSF} * \text{RoofSlopeFactor}/100 \quad (\text{fig. } 4.18)$$

RoofingPlanSF and *RoofSlopeFactor* are the names of the cells that contain the take-off inputs.

Figure 4.18. Roofing squares calculation

	Squares	▼		f_x	=RoofingPlanSF*RoofSlopeFactor/100	
	A			B	C	D
1	**Roofing Detail Sheet**					
2	**Take-off Inputs**					
3	Plan SF of Roof			1704		
4	Slope (enter the rise only)			6		
5	Fascia LF			180		
6	Starter LF			180		
7	Ridge LF			34		
8	Hip LF (A)			68		
9						
10						
11	**Calculated**					
12	Slope Factor			1.118033989		
13	Squares			19.05129917		

Roundup Function

The result, 19.05129917, is 19 full bundles, plus one partial bundle of squares. You could use the **Roundup** function to adjust the total to full bundles but rounding up by squares, rather than bundles, will be more accurate. The shingles used in the example are packaged three bundles to a square. Excel® can use this information to get a more accurate estimate. Follow these steps to round up to ⅓ of a square:

1. Multiply 19.05129917 × 3.

2. Round up to the nearest whole number.

3. Divide the result by 3.

   The formula looks like this:

   =ROUNDUP(RoofingPlanSF*RoofSlopeFactor /100*3,0)/3.

The result is 19.33 squares. The Roundup function can be used to calculate any similar fractional quantity. For example, builders typically purchase concrete in ¼ CY increments. The process for rounding up to ¼ CY is as follows:

1. Multiply the concrete thickness by the width by the length.

2. Divide by 27 (since there are 27 cu. ft. per CY).

3. Multiply the result by 4 and round up to a whole number.

4. Divide that number by 4. The round-up remainder is to the nearest ¼ CY.

Here is the Excel® formula:

=ROUNDUP((Thick″/12*Width′*Length′/27)*4,0)/4

The Roofing Database

Databases are great places for storing formulas because Excel® can automatically look up the appropriate formula based on a material selection. For example, when you select Drip Edge and add it to the detail sheet, VLOOKUP retrieves the corresponding drip edge formula from the database. When you populate the detail sheet with material or labor descriptions from a database, Excel® retrieves not only the corresponding units and unit costs, but also the appropriate formulas. The spreadsheet will then display the correct quantity of each item. This shaves hours off your estimating time.

Naming Cells

The *RoofingDetail* sheet contains four sections: Take-off Inputs, Calculated, Roofing, and Labor. We'll write formulas to determine the quantities.

Your quantity calculation formulas will reference cells in the Take-off Inputs and Calculated subsections of the *RoofingDetail* worksheet, so if you have not already done so, name all of the cells that contain the take-off inputs and the calculated values (B3:B8 and B12:B15). The following names are used in the examples: *RoofSlope, RoofingFascia, RoofingStarter, RoofingRidge, RoofingHipLF, Squares, RoofingHipFactor, RoofingActualHipLF*. Using names rather than cell locations makes the formulas easier to understand. You may adjust the sample formulas and/or cell names in your worksheets to correspond with your construction methods and materials.

Entering Formulas

Enter formulas in the QTY column of the roofing database. You may enter a default value into the QTY column if you typically only order one of each item, or you may leave the cell blank if you want to enter the quantity manually in the detail sheet.

Figure 4.19 shows a sample of the roofing database with completed formulas in the QTY column. The figure shows the worksheet in **Formula View**. To turn on Formula View, press **Ctrl + `** (the control key and the accent key simultaneously). The accent key is below the **Esc** key and above the **Tab** key at the upper left of your keyboard. You can toggle between Formula and Normal views using these keys.

The formulas work in conjunction with the named cells and formulas of the Take-off Inputs and Calculated columns on the *RoofingDetail* sheet. For example, you already have determined the number of roofing squares and named the cell that contains that value *Squares*. In the database, enter the formula **=Squares** next to any of the roofing shingle types. By entering this formula, you ensure that Excel® will calculate roofing material in squares, excluding starter strip, hip cap, or ridge cap shingles.

Drip-Edge Metal

Drip-edge Metal is available in 10′ lengths and is applied with 2 to 3″ of overlap. Your formula must also account for waste.

- Divide the lineal feet (LF) of the roof perimeter (the take-off input located in the *RoofingFascia* cell, B5) by 9.5.
- Round up to the nearest whole piece of drip edge using the following formula:

=ROUNDUP(RoofingFascia/9.5,0)

Figure 4.19. Roofing database with formulas

	A	B	C	D
1	**Roofing Material Database**			
2	**Description**	**QTY**	**Unit**	**$/Unit**
3	Asphalt Shingles - 25 yr. 3 Tab	=Squares	SQ (T)	30
4	Asphalt Shingles - 30 yr. 3 Tab	=Squares	SQ (T)	38
5	Asphalt Shingles - 30 yr. Architectural	=Squares	SQ (T)	48.43
6	Cedar Shakes #1 medium handsplits	=Squares	SQ (T)	110
7	Cedar Shingles #1	=Squares	SQ (T)	135
8	Metal Roofing	=Squares	SQ (T)	90
9	Eagle Tile	=Squares	SQ (T)	105
10	--			
11	Deliver & Stock shingles	=SUMIF(RoofingUnits,"SQ (T)",RoofingQTYs)	SQ	3.5
12	Drip Edge-metal	=ROUNDUP(RoofingFascia/9.5,0)	Ea	3.9
13	15# Tar Paper (Asphalt-impregnated Felt)	=ROUNDUP(Squares/4,0)	4-Sq Roll	10.95
14	30# Tar Paper (Asphalt-impregnated Felt)	=ROUNDUP(Squares/2,0)	2-Sq Roll	10.95
15	Ice and Water Shield - 3' x 75'	=ROUNDUP(RoofingStarter/75,0)	2-Sq Roll	77
16	Starter Strip	=ROUNDUP(RoofingStarter/240*3,0)/3	SQ (T)	38
17	Hip Cap	=ROUNDUP(RoofingActualHipLF/100*3,0)/3	SQ (T)	38
18	Ridge Cap	=ROUNDUP(RoofingRidge/100*3,0)/3	SQ (T)	38
19	Architectural Hip Cap	=ROUNDUP(RoofingHipLF/30,0)	Box	30
20	Architectural Ridge Cap	=ROUNDUP(RoofingRidge/30,0)	Box	30
21	Ridge Vents		Lf	2.3
22	Turtle-back Vents 10 x 14 (1 SF net area)		Ea	15
23	Step Flashing 3 x 4 x 7 (22' / Bndl.)		Bndl of 50	12.95
24	18" x 10' Valley Flashing		Ea	5.5
25	L Metal 4" x 4" x 10'		Ea	4.79
26	Plastic Caps 1/2 lb./sq	=ROUNDUP(Squares/25,0)	Pail	34.95
27	Simplex Nails-1/2 lb./sq	=ROUNDUP(Squares/2,0)	Lb	0.85
28	Roofing Nails - 1-1/4" - 2 lb./sq	=ROUNDUP(DeliverStockShingles*2,0)	Lb	0.62
29	Coil Roofing Nails 1" - 1-1/4"	=ROUNDUP(DeliverStockShingles/25,0)	Pail	37.95
30	8d galvanized box nails		Lb	0.95
31	6d galvanized box nails		Lb	0.95
32	4d galvanized box nails		Lb	0.95
33				
34	Labor	=DeliverStockShingles	SQ	38
35				

Asphalt-Impregnated Felt

Asphalt-impregnated felt is available in 15#/SQ rolls (4 squares per roll) and 30#/SQ rolls (2 squares per roll).

- Divide the number of squares by 4 (for 15# felt) or 2 (for 30# felt).
- Round up to the nearest whole number to calculate the number of rolls, using one of the following formulas:

=ROUNDUP(Squares/4,0)

(15#/sq. roll)

or

=ROUNDUP(Squares/2,0)

(30#/sq. roll)

- For Ice & Water Shield®, divide the LF of eaves (*RoofingStarter*) by the LF in the roll and round up to the nearest whole roll using the following formula:

=Roundup(RoofingStarter/75,0)

Ice & Water Shield® replace felt at the eaves. Therefore, deduct an equivalent amount of 15# or 30# felt when using Ice & Water Shield®. Starter strip requires 1 square of shingles for every 240 LF of eave. Round up to the nearest ⅓ of a square which is the nearest whole bundle. The formula is as follows:

=ROUNDUP(RoofingStarter/240*3,0)/3

Hip Cap

Calculate the hip cap by dividing the LF of hip by 100 LF per SQ. Construction plans do not show the hip length so you must calculate it similarly to the way you calculate squares of roofing. Use a hip factor instead of slope factor as follows:

=SQRT(RoofSlope^2+288)/12″

Figure 4.20 shows the relationship of the common rafter to the hip rafter. For every 12″ of run on the common rafter, the hip rafter has a run of 17″. Therefore, use a base number of 12 to find the slope factor and a base number of 17 to find the hip factor. The equation that follows determines the length of the run of the hip rafter.

$$= \sqrt{12^2 + 12^2} = \sqrt{288} = 1697 \approx 17$$

To find a hip factor multiplier to use in determining the true length of a hip, use 17 (the base measurement for a hip) and 6 (the rise of the hip):

$$c = \sqrt{a^2 + b^2} = \sqrt{6^2 + 17^2} = \sqrt{36″ + 289″} = \sqrt{325″} = 18″ = 1.5′$$

For a 6/12 slope, the hip factor is 1.5. In other words, to determine the length of the hip rafter, multiply the run plus the overhang of the common rafter (A or B, fig. 4.21) by the hip factor. For example, you determine the hip measurement for a roof with a span of 24′ and a 1′ overhang as follows: 13′ × 1.5′ = 19.5′. (The run is ½ the span, or 12′, plus the 1′ overhang, which equals 13′.)

Figure 4.20. Hip and common comparison

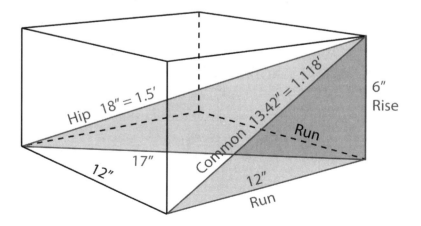

Figure 4.21. Hip and hip measurement

Ridge Cap

To determine the ridge cap, divide the ridge length by 100 LF per SQ and round up the result to the nearest bundle. The formula is as follows:

$$=ROUNDUP(RoofingRidge/100*3,0)/3$$

One box of architectural hip and ridge cap typically covers about 30 LF of hip or ridge. Plastic caps come in pails that cover about 25 squares of felt paper. Therefore, a formula for estimating cap is

$$=ROUNDUP(Squares/25,0)$$

Shingle Quantity

Look at cell B20 under QTY in the roofing detail sheet, Deliver & Stock shingles (fig. 4.22). Deliver & Stock shingles includes the squares of shingles as determined

Figure 4.22. Roofing detail sheet

	A	B	C	D	E
1	**Roofing Detail Sheet**				
2	**Take-off Inputs**				
3	Plan SF of Roof	1704			
4	Slope (enter the rise only)	6			
5	Fascia LF	180			
6	Starter LF	180			
7	Ridge LF	34			
8	Hip LF (A)	68			
9					
10					
11	**Calculated**				
12	Slope Factor	1.118			
13	Squares	19.33			
14	Hip Factor	1.500			
15	Actual Hip LF	102		**RoofingQTYs**	
16					
17	**Roofing**			**RoofingUnits**	
18	**Description**	**QTY**	**Unit**	**$/Unit**	**Sub-Total $**
19	Asphalt Shingles - 30 yr. 3 Tab	19.33	SQ (T)	$ 38.00	$ 734.67
20	Deliver & Stock shingles	22.33	SQ	$ 3.50	$ 78.17
21	Drip Edge-metal	19.00	Ea	$ 3.90	$ 74.10
22	15# Tar Paper (Asphalt-impregnated Felt)	5.00	4-Sq Roll	$ 10.95	$ 54.75
23	Starter Strip	1.00	SQ (T)	$ 38.00	$ 38.00
24	Hip Cap	1.33	SQ (T)	$ 38.00	$ 50.67
25	Ridge Cap	0.67	SQ (T)	$ 38.00	$ 25.33
26	Ridge Vents	0.00	Lf	$ 2.30	$ -
27	Step Flashing 3 x 4 x 7 (22' / Bndl.)	0.00	Bndl of 50	$ 12.95	$ -
28	L Metal 4" x 4" x 10'	0.00	Ea	$ 4.79	$ -
29	Plastic Caps 1/2 lb./sq	1.00	Pail	$ 34.95	$ 34.95
30	Coil Roofing Nails 1" - 1-1/4"	1.00	Pail	$ 37.95	$ 37.95
31	Labor	22.33	SQ	$ 38.00	$ 848.67
32				#N/A	$ -
33					
34				Total Cost:	$ 1,977.25

in the previous calculation, plus the shingles from the starter strip, hip, and ridge estimates. To calculate quantity, you must add values from the QTY column based on their description in the Unit column. I have designated the relevant quantities with a "(T)" in the unit column of the database (meaning the unit is in squares and the quantity of squares adds to the total of shingles to be delivered). You can choose any unique identifier.

Use the **SUMIF** function to total values in a column based on criteria from another column as follows:

- Name the QTY and Unit columns (excluding their headings). I named them, respectively, *RoofingQTYs* and *RoofingUnits*.
- Name the cell in the QTY column and the "Deliver and Stock shingles" row (cell B11, *RoofingMatDB* sheet), *DeliverStockShingles*.
- In the QTY column of the roofing material database (*RoofingMatDB*) next to "Deliver and Stock shingles" (**B11**), enter the following formula:

$$\text{=SUMIF(RoofingUnits,"SQ (T)",RoofingQTYs)}$$

This tells Excel® that if the *RoofingUnits* range (column C under Units on the roofing detail sheet) contains the value "SQ (T)," then sum the corresponding quantities in the *RoofingQTYs* range (column B under QTY on the *RoofingDetail* sheet).

Nails

Use the *DeliverStockShingles* value to determine the quantity of roofing nails to use for the shingles. The formulas for determining roofing nail quantities are as follows:

$$\text{=ROUNDUP(DeliverStockShingles*2,0)}$$

$$\text{=ROUNDUP(DeliverStockShingles/25,0)}$$

If you use loose nails, you need about 2# of 1¼″ roofing nails for each square of shingles. If you use 1¼″ coil roofing nails, one box will cover about 25 squares of shingles.

Cleaning Up #N/A Error Values

The error value #N/A in a cell indicates another cell or cells with the VLOOKUP arguments has the wrong information or is empty. Although the #N/A error helps you find mistakes in your spreadsheet, it also will carry over to the Total. You can solve this problem one of two ways. Both use the IF function.

The formula for a total or subtotal of a line item usually is QTY × Unit Cost (=QTY*$/Unit). If you enter the VLOOKUP formula but the description is missing, Excel® will return a #N/A error value. To avoid this error, write the formula in the total or subtotal column in one of two ways:

$$\text{=IF(ISERROR(the cell address for the \$/Unit),0,QTY*\$/Unit)}$$

or

$$\text{=IF(ISERROR(the cell address for the \$/Unit),"",QTY*\$/Unit)}$$

The first formula tells Excel®

1. If the value in the $/Unit column is an error, enter a 0 in the total or subtotal column.

2. If the value in the $/Unit column is not an error, enter the product of the value in the QTY column and the value in the $/Unit column (Quantity × Unit Cost).

The second formula tells Excel® that if there is an error value in the $/Unit column, leave the total or subtotal columns blank, as designated by quotation marks "" in the formula.

The second method of eliminating #N/A error values is testing whether the description column has data in the cell. Use one of the following two formulas:

=IF(ISBLANK(the cell address for the Description),0,QTY*$/Unit)

or

=IF(ISBLANK(the cell address for the Description), "",QTY*$/Unit)

The MATCH function

You can use the **MATCH** function to determine labor costs when these costs depend on the scope of work. With roofing, for example, the labor rate per square of shingles depends on the slope of the roof (fig. 4.23). In this example, the labor rate for an asphalt-shingle 30-yr. architectural roof with either a 5/12 or a 6/12 slope is $43.

The roofing labor database worksheet tab is *RoofingLaborDB*. Cells C3:G3 of the database contain various roofing slopes. These cells are named *RoofingSlopes*.

The roofing detail worksheet includes a section at the bottom for labor (fig. 4.24). You can create a VLOOKUP formula (cell E38 of the *RoofingDetail* sheet) to return the Labor cost per SQ for a 6/12 roof slope as follows:

=VLOOKUP(A37,RoofingLaborDB,5,FALSE)

This formula works fine as long as the roof slope does not change. However, if you changed the roof to a 10/12 slope, for example, the column index number in the VLOOKUP formula that references the *RoofingLaborDB* would need to be 3 instead of 5. To automatically change the column index number when the value of the slope is changed, you can use the MATCH function.

Figure 4.23. Roofing labor database

RoofingLaborDB	▼	*fx*	Asphalt Shingles - 25 yr. 3 Tab				
	A	B	C	D	E	F	G
1							
2		Labor Cost Per Square & Slope		Slope			
3			12	10	8	6	4
4		Asphalt Shingles - 25 yr. 3 Tab	$ 60.00	$ 55.00	$ 50.00	$ 40.00	$ 35.00
5		Asphalt Shingles - 30 yr. 3 Tab	$ 60.00	$ 55.00	$ 50.00	$ 40.00	$ 35.00
6		Asphalt Shingles - 30 yr. Architectural	$ 65.00	$ 58.00	$ 53.00	$ 43.00	$ 38.00
7		Cedar Shakes #1 medium handsplits	$ 90.00	$ 85.00	$ 75.00	$ 65.00	$ 60.00
8		Cedar Shingles #1	$ 90.00	$ 85.00	$ 75.00	$ 65.00	$ 60.00
9		Metal Roofing	$ 90.00	$ 85.00	$ 80.00	$ 70.00	$ 70.00
10		Eagle Tile	$ 90.00	$ 85.00	$ 75.00	$ 65.00	$ 60.00

Figure 4.24. Roofing labor on detail sheet

| E37 | | | *fx* =VLOOKUP(A37,RoofingLaborDB,5,FALSE) |

	A	B	C	D	E	F
1	**Roofing Detail Sheet**					
2	**Take-off Inputs**					
3	Plan SF of Roof	1704				
4	Slope (enter the rise only)	6				
5	Fascia LF	180				
6	Starter LF	180				
7	Ridge LF	34				
8	Hip LF (A)	68				
9						
10						
11	**Calculated**					
12	Slope Factor	1.118				
13	Squares	19.33				
14	Hip Factor	1.500				
15	Actual Hip LF	102				
16						
17	**Roofing**					
18	**Description**	**QTY**	**Unit**	**$/Unit**	**Sub-Total $**	
19	Asphalt Shingles - 30 yr. 3 Tab	19.33	SQ (T)	$ 38.00	$ 734.67	
20	Deliver & Stock shingles	22.33	SQ	$ 3.50	$ 78.17	
21	Drip Edge-metal	19.00	Ea	$ 3.90	$ 74.10	
22	15# Tar Paper (Asphalt-impregnated Felt)	5.00	4-Sq Roll	$ 10.95	$ 54.75	
23	Starter Strip	1.00	SQ (T)	$ 38.00	$ 38.00	
24	Hip Cap	1.33	SQ (T)	$ 38.00	$ 50.67	
25	Ridge Cap	0.67	SQ (T)	$ 38.00	$ 25.33	
26	Ridge Vents	40.00	Lf	$ 2.30	$ 92.00	
27	Step Flashing 3 x 4 x 7 (22' / Bndl.)	2.00	Bndl of 50	$ 12.95	$ 25.90	
28	L Metal 4" x 4" x 10'	1.00	Ea	$ 4.79	$ 4.79	
29	Plastic Caps 1/2 lb./sq	1.00	Pail	$ 34.95	$ 34.95	
30	Coil Roofing Nails 1" - 1-1/4"	1.00	Pail	$ 37.95	$ 37.95	
31					$ -	
32				Subtotal Material Cost:	$ 1,251.27	
33				☑ TAX	$ 90.72	
34				Total Material Cost:	$ 1,341.99	
35	**Labor**					
36	**Description**	**Slope**	**QTY**	**Unit**	**$/Unit**	**Total**
37	Asphalt Shingles - 30 yr. 3 Tab	6	22.33	SQ (T)	40.00	$ 893.33
38						

The MATCH function retrieves information by finding a position within a list. For example, in cell I3 of figure 4.24 the MATCH function returns the position of 4 (slope of 6) in the *RoofingSlopes* list. The syntax for the MATCH function is as follows:

MATCH(lookup_value,lookup_array,[match_type])

In our specific case the formula is:

=MATCH(RoofingSlope1,RoofingSlopes,-1)

RoofingSlope1 is the value of the roofing slope (cell B4 of the *RoofingDetail* sheet). *RoofingSlopes* is the name of the range that contains the list of slopes (cells C3:G3 on the *RoofingLaborDB* sheet). Microsoft®'s Help explains how MATCH works as follows:

If the MATCH type is

a. 1 or omitted, MATCH finds the largest value that is less than or equal to lookup_value. The values in the lookup_array argument must be placed in ascending order, for example, ...−2, −1, 0, 1, 2 ..., A-Z, FALSE, TRUE.

b. 0, MATCH finds the first value that is exactly equal to lookup_value. The values in the lookup_array argument can be in any order.

c. −1, MATCH finds the smallest value that is greater than or equal to lookup_value.

The values in the lookup_array argument must be placed in descending order, for example, TRUE, FALSE, Z-A, ... 2, 1, 0, −1, −2 ..., and so on.

If you use 0 as the MATCH type, the value of the slope (*RoofingSlope1*) would need to match one of the values in the list. If the *RoofingSlope1* value is 5, 7, 9, 11 (5/12, 7/12, 9/12, 11/12), Excel® returns an error. Using −1 as the MATCH type avoids the error. It allows numbers between columns, so to speak, to round up to the next highest number on the chart (the smallest value that is greater than or equal to the lookup value).

You can insert, or nest, the MATCH function in the VLOOKUP formula in place of the column index number to automatically display the labor cost per square for the selected type of roofing and a specific slope. In the example shown in cell I3 of figure 4.25, using the MATCH function to find a 6/12 slope replaces the column index number in the VLOOKUP formula with a 4. But the column index number for a 6/12 slope in the VLOOKUP formula for the *RoofinglaborDB* should be 5, not 4.

RoofingSlopes (C3:G3) is offset from the named range *RoofingLaborDB* (B4:G4) by 1 column. To correct for the misalignment, either add 1 to the value of the MATCH function or change the named range *RoofingSlopes* from C3:G3 to B3:G3.

The VLOOKUP function to find the labor cost per square (cell E37, fig. 4.23) is:

$$\text{=VLOOKUP(A37,RoofingLaborDB,5,FALSE)}$$

If you replace the column index location with the MATCH function, MATCH will automatically adjust unit costs depending on the slope of the roof. The formula is

$$\text{=VLOOKUP(A37,RoofingLaborDB,MATCH(RoofingSlope1,}$$
$$\text{RoofingSlopes,−1)+1,FALSE)}$$

Notice the addition of +1 at the end of the MATCH function to correct for the offset of the *RoofingSlopes* and the *RoofingLaborDB* named ranges.

Figure 4.25. The MATCH function in a formula

I3		f_x =MATCH(RoofingSlope1,RoofingSlopes,-1)						
A	B	C	D	E	F	G	H	I
1								
2	Labor Cost Per Square & Slope			Slope				
3		12	10	8	6	4		4
4	Asphalt Shingles - 25 yr. 3 Tab	$ 60.00	$ 55.00	$ 50.00	$ 40.00	$ 35.00		
5	Asphalt Shingles - 30 yr. 3 Tab	$ 60.00	$ 55.00	$ 50.00	$ 40.00	$ 35.00		
6	Asphalt Shingles - 30 yr. Architectural	$ 65.00	$ 58.00	$ 53.00	$ 43.00	$ 38.00		
7	Cedar Shakes #1 medium handsplits	$ 90.00	$ 85.00	$ 75.00	$ 65.00	$ 60.00		
8	Cedar Shingles #1	$ 90.00	$ 85.00	$ 75.00	$ 65.00	$ 60.00		
9	Metal Roofing	$ 90.00	$ 85.00	$ 80.00	$ 70.00	$ 70.00		
10	Eagle Tile	$ 90.00	$ 85.00	$ 75.00	$ 65.00	$ 60.00		

"What If?" Scenarios

The completed roofing detail sheet allows you to quickly and accurately create thorough take-offs, cost estimates, and bids. When your customer asks, "What if we change from a 6/12 to an 8/12 roof slope?" and "How much will it cost?" you will only need to change the *RoofSlope* value to 8 to answer them. Excel® will automatically adjust quantities and total costs for the 8/12 slope (fig. 4.26).

Protecting Cells from Accidental Erasure

Once you create a spreadsheet with all the necessary formulas, formats, and features to enable quick and accurate take-offs, you can protect the sheet from accidental erasures or other changes. You may wish to do this if more than one person will be entering data into a spreadsheet.

When you protect a sheet, each cell's security format is locked. Therefore, if you want to allow access to a cell, you must unlock that cell before protecting the worksheet. Protecting a worksheet is a two-step process:

First, you must unprotect the cells that you want to allow changes to. Second, you must protect the sheet. To change the protection setting for a cell or cells:

- Right click the selection and click Format Cells from the pop-up menu (fig. 4.27). (Another method is to click the arrow at the lower right of the **Font**, **Alignment**, or **Number** sections of the **Home** tab of the ribbon to reveal the **Protection** tab.)
- Click the **Protection** tab (fig. 4.28). You will notice that the cells are locked by default. To unlock a cell, uncheck the **Locked** box.

Note the **Hidden** option. Checking that box allows you to hide formulas on a protected sheet. Typically, you do not want to hide formulas. Leaving them visible even if the cells are locked and protected allows users to view the formulas, verify their accuracy, and understand how Excel® is processing the data.

After you unlock or lock cells as desired, you can protect the sheet as follows:

- Click the **Review** tab of the ribbon and **Protect Sheet** (fig. 4.29).

Excel® displays a number of options including password protection. Although the default settings are usually adequate, you may allow users additional access by selecting various options.

Figure 4.26. Roofing labor database and slopes>

	Labor Cost Per Square & Slope		Slope						4
			12	10	8	6	4		
	Asphalt Shingles - 25 yr. 3 Tab		$ 60.00	$ 55.00	$ 50.00	$ 40.00	$ 35.00		RoofingSlopes
	Asphalt Shingles - 30 yr. 3 Tab		$ 60.00	$ 55.00	$ 50.00	$ 40.00	$ 35.00		
	Asphalt Shingles - 30 yr. Architectural		$ 65.00	$ 58.00	$ 53.00	$ 43.00	$ 38.00		
	Cedar Shakes #1 medium handsplits		$ 90.00	$ 85.00	$ 75.00	$ 65.00	$ 60.00		
	Cedar Shingles #1		$ 90.00	$ 85.00	$ 75.00	$ 65.00	$ 60.00		
	Metal Roofing		$ 90.00	$ 85.00	$ 80.00	$ 70.00	$ 70.00		
	Eagle Tile		$ 90.00	$ 85.00	$ 75.00	$ 65.00	$ 60.00		
	RoofinglaborDB								

Figure 4.27. Changing cell formats

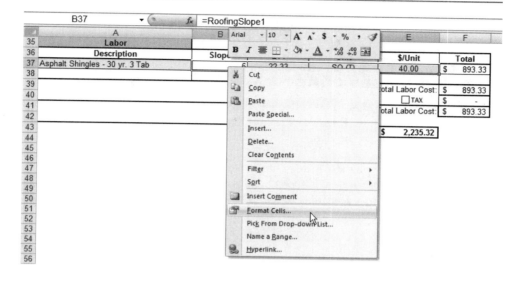

Figure 4.28. Changing cell protection

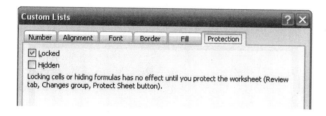

Figure 4.29. Protecting a worksheet

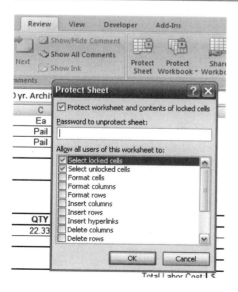

Unless you have proprietary formulas or information you need to preserve, I do not recommend using password protection because if you forget the password, you will not be able to access the locked cells. If you choose to use a password, however, write it down and keep it in a safe location.

Databases and Detail Sheets: The Workhorses of Accurate Estimating

Detail sheets contain specific information about items included in an estimate. They also contain formulas that automate most of the estimating process. You can add formulas to the detail sheets that enable Excel® to look up information from databases.

When suppliers or trade contractors change their prices, you can change your databases accordingly. Those changes will automatically update the detail sheet. Also, if you store formulas in the QTY column of the database, you can automate quantity calculations and save time.

Saving Time with Hyperlinks

5

You probably have used software programs that did not interface with other programs. After finishing a task in one program or application, you had to copy and paste the information into another one. Worse yet is retyping information that can't be copied. Copying, pasting, and retyping information is time-consuming and frustrating. Most builders and remodelers don't have time to waste.

In this chapter you will learn how to link your worksheets with other spreadsheets, workbooks, and files. Use your completed Chapter 4 workbook, which includes the following worksheets:

1. Cost breakdown summary sheet (*CostBreakdownSummary*)
2. Roofing detail (*RoofingDetail*)
3. Roofing database (2 worksheets—*RoofingMatDB* and *RoofingLaborDB*)

In developing your own estimating system, you could place the database in a different workbook from the cost breakdown summary and roofing detail sheets. You can decide which option works best for you and your company.

Linking

Linking replicates the value of one cell to another without copying, pasting, or retyping. The value in the linked (destination) cell mirrors the value in the original (source) cell. Look at E43 in figure 5.1. Wouldn't it save time if the Total Material and Labor Cost calculations on the roofing detail sheet automatically updated the roofing total on the cost breakdown summary sheet? Using the linking function in Excel®, your spreadsheets will do just that. For example, when a roofing quantity or price changes, the new roofing total will instantly appear in the "Est. Cost" or "Bid" column next to Roofing (cell F44) on the cost breakdown summary sheet.

Figure 5.1. Roofing detail sheet

	A	B	C	D	E	F
1	Roofing Detail Sheet					
2	Take-off Inputs					
3	Plan SF of Roof	1704				
4	Slope (enter the rise only)	6				
5	Fascia LF	180				
6	Starter LF	180				
7	Ridge LF	34				
8	Hip LF (A)	68				
9						
10						
11	Calculated					
12	Slope Factor	1.118				
13	Squares	19.33				
14	Hip Factor	1.500				
15	Actual Hip LF	102				
16						
17	Roofing					
18	Description	QTY	Unit	$/Unit	Sub-Total $	
19	Asphalt Shingles - 30 yr. 3 Tab	19.33	SQ (T)	$ 38.00	$ 734.67	
20	Deliver & Stock shingles	22.33	SQ	$ 3.50	$ 78.17	
21	Drip Edge-metal	19.00	Ea	$ 3.90	$ 74.10	
22	15# Tar Paper (Asphalt-impregnated Felt)	5.00	4-Sq Roll	$ 10.95	$ 54.75	
23	Starter Strip	1.00	SQ (T)	$ 38.00	$ 38.00	
24	Hip Cap	1.33	SQ (T)	$ 38.00	$ 50.67	
25	Ridge Cap	0.67	SQ (T)	$ 38.00	$ 25.33	
26	Ridge Vents	40.00	Lf	$ 2.30	$ 92.00	
27	Step Flashing 3 x 4 x 7 (22' / Bndl.)	2.00	Bndl of 50	$ 12.95	$ 25.90	
28	L Metal 4" x 4" x 10'	1.00	Ea	$ 4.79	$ 4.79	
29	Plastic Caps 1/2 lb./sq	1.00	Pail	$ 34.95	$ 34.95	
30	Coil Roofing Nails 1" - 1-1/4"	1.00	Pail	$ 37.95	$ 37.95	
31					$ -	
32				Subtotal Material Cost:	$ 1,251.27	
33				☑ TAX	$ 90.72	
34				Total Material Cost:	$ 1,341.99	
35	Labor					
36	Description	Slope	QTY	Unit	$/Unit	Total
37	Asphalt Shingles - 30 yr. 3 Tab	6	22.33	SQ (T)	40.00	$ 893.33
38					Subtotal Labor Cost:	$ 893.33
39					☐ TAX	$ -
40					Total Labor Cost:	$ 893.33
41						
42						
43				Total Material and Labor Costs	$ 2,235.32	
44						

You can create a link using one of two methods. The first one is the easiest.

- Select cell F44 on the cost breakdown summary sheet.
- Enter the = sign.
- Click the *RoofingDetail* sheet tab.
- Click the cell that contains the value for the roofing total, or the source, E44. Excel® automatically inserts the cell address into the formula.
- Press **Enter** to complete the formula (fig. 5.2).

Test your link by changing a quantity or an item description on the roofing detail sheet and checking the cost breakdown summary sheet to verify the update.

You also could name the cell that contains the roofing total (E43 on the roofing detail sheet) and use it for the linked reference (fig. 5.3).

Figure 5.2. Creating a link

	B	C	E	F	G
			F44	=RoofingDetail!E43	
42			Electrical	$ -	
43			Light Fixture Allowance	$ -	
44			Roofing	$ 2,235.32	
45			Insulation	$ -	
46			Drywall	$ -	

Figure 5.3. Using a cell name in a link

	B	C	E	F
			F44	=RoofingTotal
42			Electrical	$ -
43			Light Fixture Allowance	$ -
44			Roofing	$ 3,190.99
45			Insulation	$ -
46			Drywall	$ -

You could also

- **Copy** the source information. (In this case, on the roofing detail sheet, copy E43).
- Navigate to the cost breakdown summary sheet and select F44.
- Click the **Paste menu** on the **Home** tab of the ribbon.
- Click **Paste Special.**
- Click **Paste Link.**

Adding Hyperlinks

When you have only a few worksheets in your workbook, you can move easily from one sheet to another by clicking the tabs at the bottom of the workbook. However, as you add more sheets to your estimating program, navigating by clicking tabs becomes slow and cumbersome. Adding hyperlinks to your worksheets will solve this problem.

You can add hyperlinks to text and graphics and then place these links in other locations. When you click them, Excel® automatically selects the cell the hyperlink references. Follow this example:

- Select cell E44 on the cost breakdown summary sheet.
- Click the **Insert** tab on the ribbon
- Click **Hyperlink.** The Edit Hyperlink dialog box will appear (fig. 5.4).
- Click the **Place in this Document** button.
- Click *RoofingTotal* under the Defined Names section.
- Click **OK** to complete the hyperlink.

Figure 5.4. Hyperlinking to the RoofingTotal cell

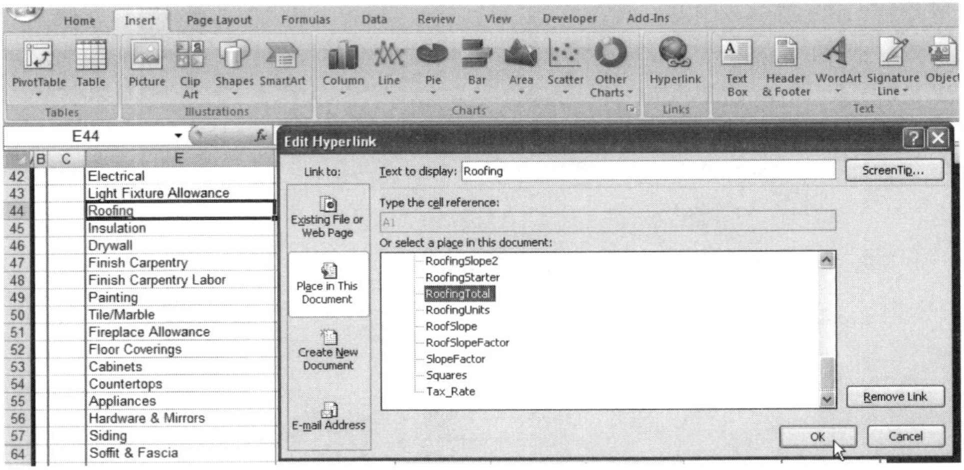

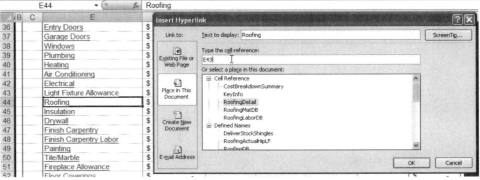

Now, whenever a user clicks the hyperlink (cell E44 on the cost breakdown summary sheet), Excel® will navigate to the *RoofingTotal*.

Hyperlinks using a named reference rather than a cell reference (=*Roofing Detail!E43*) will remind you what you are linking to and what the cell contains.

A hyperlink also can refer to a specific cell address (fig. 5.5). However, if you insert or delete cells in the worksheet, the link does not move with the original cell. For example, if you insert or delete a row above the roofing total cost (E43), the hyperlink will not move with the roofing total cost.

You also can hyperlink to other files. If you had a document with specifications and you want to access the file from Excel®

- Follow the steps in the previous section to select a location for the hyperlink.
- Right click the cell where you want to place the hyperlink.
- Click **hyperlink**.

Figure 5.5. Hyperlinking to a cell reference

- Choose **Existing File** or **Web Page**.
- Browse to the file you want to link to. Excel® enters the URL (*uniform resource locator*—the Web address).

If you need information from the *NAHB Chart of Accounts*, for example

- Click the **Existing File** or **Web Page** box.
- Enter *http://www.nahb.org/chart* into the **Address:** box or click the **Browse Internet** button to navigate to the desired location (fig. 5.6).
- Click **OK**.

Hyperlinks help you navigate quickly to another cell in the same worksheet, to a cell in a different worksheet, to another file, or to an Internet source. Linking connects all of your work into an organized package so you can easily share and summarize information.

Figure 5.6. Hyperlinking to a Web page

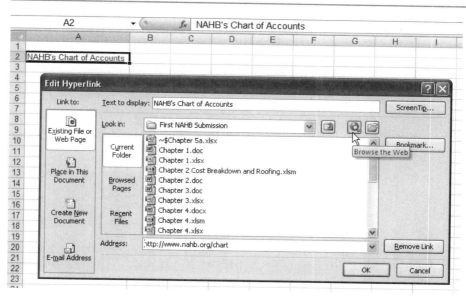

Letting Formulas and Functions Do the Work

6

Now that you have seen the remarkable power that functions such as VLOOKUP, MATCH, IF, SUM, and NOW add to the formulas that automate your spreadsheets, you are ready to learn additional functions and formulas to increase the capability of your spreadsheets and enable you to complete your estimates in even less time.

This chapter explores in more depth how to use formulas and functions to further automate calculating materials quantities and permit fees. By adapting the sample formulas, you will be able to calculate quantities for most items in your estimating spreadsheets.

You will learn applications to improve spreadsheet estimating and new ways to apply functions such as ROUNDUP, IF, and VLOOKUP.

You will set up detailed estimating worksheets for

- concrete
- rebar
- permit fees
- connection and impact fees

While creating these detailed estimating sheets, you will learn to write formulas that will greatly improve the speed and accuracy of your estimates in three specific areas:

- concrete quantities
- rebar quantities
- waste factoring

In addition, you will learn how to format your spreadsheets to improve readability and how to annotate your spreadsheets with comments.

Estimating Concrete

Although automated spreadsheets can reduce much of the drudgery of estimating, the estimator still must provide take-off values. As discussed previously, however, these values should be the only inputs needed. If you must use a calculator before inputting values, you are doing work that the computer could do for you.

For example, many estimators input the CY of concrete and let the computer multiply the CY needed by the cost per CY to calculate the total cost. But CY is not a take-off quantity; it is a calculated quantity. You can design your spreadsheets to require only take-off quantities and eliminate the intermediate step of calculating volume. This approach also preserves a record of take-off details to verify amounts, control costs, and provide information for future reference.

The Chapter 6 workbook contains several worksheets. You will learn the formulas by entering them in the *Flatwork* worksheet. The *Flatwork (Completed)* worksheet contains all of the completed formulas so you can check your work. The sample workbook combines the flatwork database and the detail sheet on one worksheet so you don't have to flip back and forth between worksheets while working through the examples.

Enter take-off values into the areas indicated by the arrows in figure 6.1. Using VLOOKUP, Excel® can find the slab thicknesses and PSI from the default values in

Figure 6.1. Flatwork detail and database

Flatwork Detail					Sub		Waste		Supplier	
Flatwork										
Description	Thick in.	Width ft.	Length ft.	SF	Labor $/SF	Labor Sub-Total	CY	PSI	$/CY	Material Sub-Total
Garage Slab										
Driveway										
Drive Approach Apron										
				0						
End			SubTotals:	0		$ -	0.00			
									Sales Tax:	$ -
									Total Material and Labor:	$ -

Take-off Values

Flatwork Labor Database							
Flatwork	Thickness "	Unit	PSI	Unit	Sub 1	Sub 2	Sub 3
Basement Slab	4	SF	3000	CY	$ 0.65	$ 0.68	$ 0.72
Slab on Grade	4	SF	3000	CY	$ 0.65	$ 0.68	$ 0.72
Porch Cap	6	SF	4000	CY	$ 0.85	$ 0.89	$ 0.94
Garage Slab	4	SF	3000	CY	$ 0.75	$ 0.79	$ 0.83
Driveway	4	SF	4000	CY	$ 0.75	$ 0.79	$ 0.83
Drive Approach Apron	4	SF	4500	CY	$ 0.80	$ 0.84	$ 0.88
Patio	4	SF	4000	CY	$ 0.75	$ 0.79	$ 0.83
Sidewalk Labor	4	SF	4000	CY	$ 0.70	$ 0.74	$ 0.77
Sidewalk Base	4	SF	Road Base	CY	$ 0.70	$ 0.74	$ 0.77
Theater Slab	4	SF	3000	CY	$ 0.85	$ 0.89	$ 0.94

Concrete Database				
PSI Concrete	Unit	Supplier 1	Supplier 2	Supplier 3
2500	CY	$ 87.70	$ 92.09	$ 96.48
3000	CY	$ 92.24	$ 96.86	$ 101.47
3500	CY	$ 95.26	$ 100.03	$ 104.79
4000	CY	$ 98.28	$ 103.20	$ 108.11
4500	CY	$ 104.33	$ 109.55	$ 114.77
5000	CY	$ 107.36	$ 112.73	$ 118.10
Road Base	CY	$ 8.40	$ 8.82	$ 9.24

the flatwork labor database. When you select the flatwork item and enter the width and length, Excel® will look up or calculate all other values. Let's set up the concrete detail sheet. You may recall some of the following instructions from Chapter 4.

Before entering formulas into the cells, name the various cells and cell ranges that the formulas will use as references. Use names that will help you remember what information the cells and ranges contain. Figure 6.2 shows examples of named cells and cell ranges on the flatwork detail worksheet.

- Use **Data Validation** (in the **Data** tab on the ribbon) to enter descriptions of the flatwork items in column B (B6:B9). Allow values from a list. The source of this list is the descriptions in the flatwork labor database (B16:I26) located below the flatwork detail section on the flatwork worksheet.

Because the normal thickness of concrete slabs typically does not vary, you should store these thicknesses as default values in the database (C1:C26) and retrieve them to the "Thick in." column (C6:C9) in the Flatwork Detail using VLOOKUP.

- Enter the following formula into cell C6:

=VLOOKUP(B6,FlatworkLaborDB,2,FALSE)

- **Copy** the formula to cells C7:C9.

Figure 6.2. Named cells and cell ranges

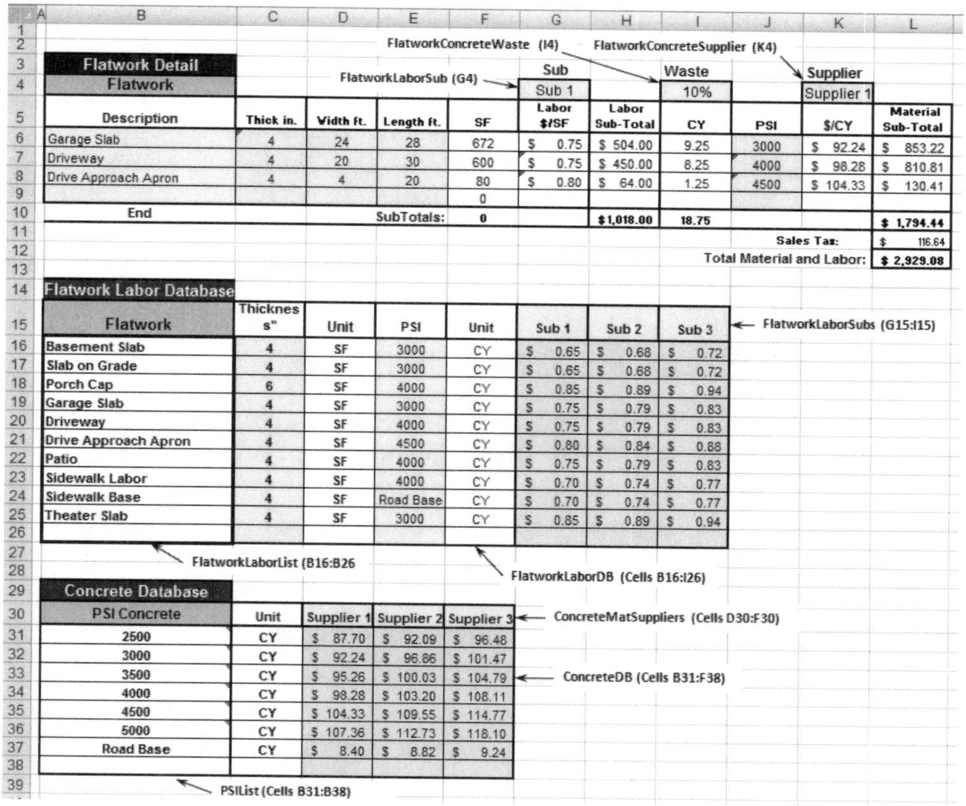

Flatwork Detail — Flatwork; Sub: Sub 1; Waste: 10%; Supplier: Supplier 1

Description	Thick in.	Vidth ft.	Length ft.	SF	Labor $/SF	Labor Sub-Total	CY	PSI	$/CY	Material Sub-Total
Garage Slab	4	24	28	672	$ 0.75	$ 504.00	9.25	3000	$ 92.24	$ 853.22
Driveway	4	20	30	600	$ 0.75	$ 450.00	8.25	4000	$ 98.28	$ 810.81
Drive Approach Apron	4	4	20	80	$ 0.80	$ 64.00	1.25	4500	$ 104.33	$ 130.41
				0						
End			SubTotals:	0		$1,018.00	18.75			$ 1,794.44
									Sales Tax:	$ 116.64
									Total Material and Labor:	$ 2,929.08

Flatwork Labor Database

Flatwork	Thickness"	Unit	PSI	Unit	Sub 1	Sub 2	Sub 3
Basement Slab	4	SF	3000	CY	$ 0.65	$ 0.68	$ 0.72
Slab on Grade	4	SF	3000	CY	$ 0.65	$ 0.68	$ 0.72
Porch Cap	6	SF	4000	CY	$ 0.85	$ 0.89	$ 0.94
Garage Slab	4	SF	3000	CY	$ 0.75	$ 0.79	$ 0.83
Driveway	4	SF	4000	CY	$ 0.75	$ 0.79	$ 0.83
Drive Approach Apron	4	SF	4500	CY	$ 0.80	$ 0.84	$ 0.88
Patio	4	SF	4000	CY	$ 0.75	$ 0.79	$ 0.83
Sidewalk Labor	4	SF	4000	CY	$ 0.70	$ 0.74	$ 0.77
Sidewalk Base	4	SF	Road Base	CY	$ 0.70	$ 0.74	$ 0.77
Theater Slab	4	SF	3000	CY	$ 0.85	$ 0.89	$ 0.94

Concrete Database

PSI Concrete	Unit	Supplier 1	Supplier 2	Supplier 3
2500	CY	$ 87.70	$ 92.09	$ 96.48
3000	CY	$ 92.24	$ 96.86	$ 101.47
3500	CY	$ 95.26	$ 100.03	$ 104.79
4000	CY	$ 98.28	$ 103.20	$ 108.11
4500	CY	$ 104.33	$ 109.55	$ 114.77
5000	CY	$ 107.36	$ 112.73	$ 118.10
Road Base	CY	$ 8.40	$ 8.82	$ 9.24

You could enter one of the following two equivalent formulas instead (to remove error values that Excel® may return):

=IF(ISBLANK(B6),"",VLOOKUP(B6,FlatworkLaborDB,2,FALSE)

The preceding formula translates as follows: "If cell B6 is blank, then leave the cell blank ("")... If not, use the VLOOKUP formula (to look up the thickness of flatwork from the flatwork labor database)."

The second equivalent formula is as follows:

=IF(ISERROR(VLOOKUP(B6,FlatworkLaborDB,2,FALSE)),"",

VLOOKUP(B6,FlatworkLaborDB,2,FALSE))

This formula states that if the VLOOKUP formula returns an error value, then the cell will be left blank ("") or the VLOOKUP formula will return the thickness value from the flatwork labor database.

- Enter the inputs for the garage slab width and length in feet (24 and 28, respectively) into cells D6 and E6.

- Enter one of three possible formulas into cell F6 to determine SF of the garage slab as in the following example:

=D6*E6

or

=IF(ISERROR(D6*E6),"",D6*E6)

or

=IF(ISBLANK(B6),"",D6*E6)

The first formula is the product of width × length. The next formula is an IF statement. It says, "If there is an error in the formula =D6*E6, then leave cell F6 blank. If not, display the product of the width × length." The third formula is similar: The IF statement checks to see whether cell B6 is blank. If it is blank, then Excel® leaves cell F6 blank. If cell B6 is not blank, then Excel® displays the product of width × length.

The labor cost per SF is a VLOOKUP formula. VLOOKUP will search the flatwork labor database according to the user's choice of trade contractor (cell G4). Using VLOOKUP and MATCH will return the proper cost.

- Enter one of the three following formulas into cell G6:

=VLOOKUP(B6,FlatworkLaborDB,MATCH
(FlatworkLaborTrade,FlatworkLaborTrades,0)+5,FALSE)

or

=IF(ISERROR(VLOOKUP(B6,FlatworkLaborDB,MATCH
(FlatworkLaborTrade,

FlatworkLaborTrades,0)+5,FALSE)),"",VLOOKUP
(B6,FlatworkLaborDB,

MATCH(FlatworkLaborTrade,FlatworkLaborTrades,0)+5,FALSE))

or

=IF(ISBLANK(B6),"",VLOOKUP(B6,FlatworkLaborDB,

MATCH(FlatworkLaborTrade,FlatworkLaborTrades,0)+5,FALSE))

Remember, the MATCH function returns the value of the position of the Lookup value within a list of items.

- Select Trade 1 as the Lookup value (the value in G4). The list of Trades is *FlatworkLaborTrades* (G15:I15).

- Add 5 to the MATCH function. (Although Trade 1 is the first item in the list, the *FlatworkLaborTrades* named range (G15:I15) is offset 5 columns from the *FlatworkLaborDB* (B16:I26). You must adjust for the offset to enable VLOOKUP to identify the proper trade contractor column.)

- Enter the following formula into cell H6 to determine the Labor Subtotal:

=F6*G6

- The formula finds the product of the SF and the cost per SF

- Use the ISERROR or ISBLANK functions to remove potential error values that would be returned to cells F6 through F9 if cells D6 through E9 were blank, as shown in the following two examples:

=IF(ISERROR(F6*G6),“ ”,F6*G6)

or

=IF(ISBLANK(G6),“ ”,F6*G6)

To find CYs of concrete, you must first determine cubic feet (CF) as follows:

- Divide the thickness in inches by 12 and multiply the result by the width and length in feet. Because there are 27 CF in a CY ($3 \times 3 \times 3 = 27$), you divide the total CF measurement by 27. The formula is

=(C6/12*D6*E6)/27

- To round up to the nearest ¼ CY, use this formula:

= ROUNDUP((C6/12*D6*E6)/27*4,0)/4

- Use the following formula to remove potential error values:

=IF(ISERROR(ROUNDUP((C6/12*D6*E6)/27*4,0)/4),“ ”, ROUNDUP((C6/12*D6*E6)/27*4,0)/4)

Adding Waste

You can include a waste factor in the CY calculations. Add it after calculating the CY but before rounding up to the nearest ¼ CY.

Enter the following formula into cell I4:

=IF(ISERROR(ROUNDUP((C6/12*D6*E6)/27* (1+FlatworkConcreteWaste)*4,0)/4),“ ”,ROUNDUP ((C6/12*D6*E6)/27*(1+FlatworkConcreteWaste)*4,0)/4)

The formula reads, “If there is an error in the CY formula, leave the cell blank. If the formula does not return an error, enter the value of the formula.” The formula takes the CY (thickness × width × length), adds the waste, and then rounds up to the nearest ¼ CY.

- Use VLOOKUP to find the default PSI value (or sack mix or bag mix) from the flatwork labor database for cells J6:J9. Or, type in a PSI value specific to the project. The basic formula to look up the default PSI value is:

=VLOOKUP(B6,FlatworkLaborDB,4,FALSE)

To avoid error values, use the following formula:

=IF(ISERROR(VLOOKUP(B6,FlatworkLaborDB,4,FALSE)),"",
(VLOOKUP(B6,FlatworkLaborDB,4,FALSE)))

The formula tells Excel® to leave the cell blank if B6 is blank or to look up the labor rate if B6 contains a value.

Concrete costs per CY (K6:K9) are based on the PSI designated in cells J6:J9, which Excel® looks up from the concrete database at the bottom of the flatwork worksheet.

- Enter the following formula in cell K6:

=VLOOKUP(J6,ConcreteDB,MATCH(FlatworkConcreteSupplier,
ConcreteMatSuppliers,0)+2,FALSE)

Notice the 2-column offset of the *ConcreteMatSuppliers* list (cells D30:F30) and the *ConcreteDB* (B31:F38) in the concrete database. The **MATCH** function in the formula contains a +2 to compensate for this offset.

- To remove potential error values, combine the previous two formulas and add the error check function (*ISERROR*) to create the following formula:

=IF(ISERROR(VLOOKUP(J6,ConcreteDB,MATCH
(FlatworkConcreteSupplier,
ConcreteMatSuppliers,0)+2,FALSE)),"",VLOOKUP(J6,ConcreteDB,
MATCH(FlatworkConcreteSupplier,ConcreteMatSuppliers,0)+2,FALSE))

This formula tells Excel® to display the value of the VLOOKUP formula. If the VLOOKUP formula returns an error, leave the cell blank. Otherwise, the VLOOKUP formula looks up the value in cell J6 (the PSI) in the *ConcreteDB* and returns the unit cost of the chosen supplier.

- Enter the following formula in cell L6 to find the material subtotal:

=I6*K6

To remove potential error values, you can use this formula:

=IF(ISERROR(I6*K6),"",I6*K6)

After you finish entering and checking the formulas in row 6 of the Flatwork Detail section, copy the formulas through L8 using the fill handle (fig. 6.3):

- Select the cells to copy, and right click the fill handle (lower right of the selected cells).
- Drag to the cells you desire.
- Select **Fill Without Formatting** from the pop-up menu.

Figure 6.3. Copying formulas using the Fill handle

	F	G	H	I	J	K	L	M	N	O
	ConcreteWaste (I4)			FlatworkConcreteSupplier (K4)						
	Sub (G4)	**Sub**		**Waste**		**Supplier**				
		Sub 1		10%		Supplier 1				
	SF	Labor $/SF	Labor Sub-Total	CY	PSI	$/CY	Material Sub-Total			
	672	$ 0.75	$ 504.00	9.25	3000	$ 92.24	$ 853.22			
	600	$ 0.75	$ 450.00	8.25	4000	$ 98.28	$ 810.81			
	80	$ 0.80	$ 64.00	1.25	4500	$ 104.33	$ 130.41			
	0									
s:	0		$1,018.00	18.75			$ 1,794.44			
						Sales Tax:	$ 116.64			
					Total Material and Labor:		$ 2,929.08			

(context menu: Copy Cells / Fill Series / Fill Formatting Only / Fill Without Formatting / Fill Days)

To review, the labor subtotal is the sum of cells H6:H9. The CY subtotal is the sum of cells I6:I9. The material subtotal is the sum of cells L6:L9.

Adding Taxes

Sales tax typically applies only to materials, although some states and localities apply sales tax to labor as well. Calculating sales tax is simple. You merely multiply the material subtotal by the sales tax rate. (*See SalesTax* worksheet in the Chapter 6 workbook for an example in which a named cell, *TaxRate*, contains the sales tax.) Material costs, tax, and labor are totaled from three cells using one of the following two formulas:

=H10+SUM(L10:L11)

or

=H10+L10+L11

Footings

Estimating footing costs is similar to estimating for flatwork and foundations and comprises the following three areas:

- labor
- concrete
- rebar and miscellaneous

Labor

For footings and foundations, calculate labor, if self-performed, on an hourly basis. When you use trade contractors to do the, footings and foundations, you normally determine labor based on LF. For footings, labor per LF typically includes forms, stakes, spreaders, tie wire, form oil, minimal hand grading, and the labor to place the concrete.

Concrete

The concrete price per CY is based on its volume and strength (PSI), as well as additives to the concrete mix, such as hot water or calcium.

Rebar and Miscellaneous

The third area to estimate, rebar and miscellaneous, includes 20' horizontal rebar, vertical j-bar hook, footing steps, and blockouts.

The following example demonstrates a sample footing database and detail sheet that calculate and summarize footing costs, including labor, concrete, rebar, and miscellaneous.

Figure 6.4 shows three databases on one worksheet: footing (*FtgDB*), rebar and miscellaneous (*FTGRebarMiscDB*), and concrete (*ConcreteDB*). You already have used the concrete database in the flatwork example. Both the footing and rebar and miscellaneous databases store typical footing sizes, including width and

Figure 6.4. Footing and concrete databases

Footing Database	QTY	Unit	Width "	Thick "	PSI	$/CY	Labor
Ribbon Footings----------							
16 x 8 Footing	1	LF	16	8	3000	$ 92.24	$ 3.50
18 x 8 Footing	1	LF	18	8	3000	$ 92.24	$ 3.50
18 x 10 Footing	1	LF	18	10	3000	$ 92.24	$ 3.50
20 x 8 Footing	1	LF	20	8	3000	$ 92.24	$ 3.50
20 x 10 Footing	1	LF	20	10	3000	$ 92.24	$ 3.60
20 x 12 Footing	1	LF	20	12	3000	$ 92.24	$ 3.70
24 x 10 Footing	1	LF	24	10	3000	$ 92.24	$ 3.80
24 x 12 Footing	1	LF	24	12	3000	$ 92.24	$ 3.90
30 x 12 Footing	1	LF	30	12	3000	$ 92.24	$ 4.00
30 x 16 Footing	1	LF	30	16	3000	$ 92.24	$ 4.10

Rebar & Misc DB			
PSI Concrete	QTY	Unit	Supplier 1
20'	31	EA	$ 9.16
30" J-Bar	12	EA	$ 1.89
48" J-Bar	12	EA	$ 2.75
4'-6" J-Bar	12	EA	$ 2.95
Step-up	1	EA	$ 25.00
Blockout	1	EA	$ 15.00
Bulkhead	1	EA	$ 15.00

Concrete Database			
PSI Concrete	Unit	Supplier 1	Supplier 2	Supplier 3
2500	CY	$ 87.70	$ 92.09	$ 96.48
3000	CY	$ 92.24	$ 96.86	$ 101.47
3500	CY	$ 95.26	$ 100.03	$ 104.79
4000	CY	$ 98.28	$ 103.20	$ 108.11
4500	CY	$ 104.33	$ 109.55	$ 114.77
5000	CY	$ 107.36	$ 112.73	$ 118.10
Road Base	CY	$ 8.40	$ 8.82	$ 9.24

thickness, units, default PSI specifications, CY prices (from the concrete database), labor prices per unit (LF), rebar, and miscellaneous items.

Figure 6.5 shows a sample footing detail sheet. In addition to footings measurements in linear feet, you need to enter other take-off values, including the number (QTY) of horizontal rebars in the footing, vertical j-bar spacing, and applicable waste factors.

The *Footing* worksheet in the Chapter 6 workbook contains the formulas and inputs for each column. Starting in the middle section (cells B9:B10) with the footing concrete and labor details description and working from left to right, here's how to recreate the worksheet:

- In the *FootingDB* worksheet, name cells B3:B14 *FtgList* to create a reference for Data Validation.
- In the *Footing* worksheet (cells B9:B10), populate the Description column by typing in items or by choosing from a drop-down menu using the **Data Validation** command in the **Data** tab of the ribbon.
- Name cells B3:B14 to create a reference for Data Validation.
- Click **Data Validation.**
- Choose **Data Validation** from the drop-down menu.
- Click the **Settings** tab.
- Choose **List** under the **Allow** menu.
- In the **Source:** box, type the name you assigned to cells B3:B14 of the footing database.
- Enter the quantity (QTY) manually or use a digital scale such as a Scale Master II®, a digitizer, or digital take-off software.

Figure 6.5. Footing detail sheet

	A	B	C	D	E	F	G	H	I
1									
2		**Footing Take-offs**							
3		QTY Horizontal Bars in FTG:	2					Total	$ 6,207.12
4		J-Bar Spacing in FTG (in.) :	24						
5		Footings Waste Factor:	5%						
6									
7		**Ftg Concrete & Labor**							
8		Description	QTY	Unit	CY	Concrete $/Unit	Concrete Total $	Labor $/Unit	Labor Total $
9		18 x 8 Footing	45	LF	1.75	$ 92.24	$ 161.42	$ 3.50	$ 564.97
10		20 x 10 Footing	220	LF	12.00	$ 92.24	$ 1,106.88	$ 3.60	$ 3,984.77
11									
12						Sales Tax	$ 82.44		
13		Tot LF Rebar	265	Tot CY	13.75	Conc. Tot	$ 1,350.74	Lab. Tot	$ 4,549.74
14									
15		**Rebar & Misc.**							
16		Description	QTY	Unit	$/Unit	Sub-Total $			
17		20'	31	EA	$ 9.16	$ 283.96			
18		30" J-Bar	12	EA	$ 1.89	$ 22.68			
19									
20		Total				$ 306.64			

- Enter the following VLOOKUP formula in cell D9. It tells Excel® to look up the unit from the footing database.

$$=VLOOKUP(B9,FtgDB,3,FALSE)$$

- Enter the following formula, which will calculate CYs of concrete, in cell E9:

$$=ROUNDUP(C9*VLOOKUP(B9,FtgDB,4,FALSE)/12*$$
$$VLOOKUP(B9,FtgDB,5,FALSE)/12/27*(1+Waste)*4,0)/4$$

Remember, the CY formula first multiplies thickness (ft.) × width (ft.) × length (ft.) to get CF, then divides by 27 to get CYs. Cell C9 is the length in feet. Excel® retrieves the formula, **VLOOKUP(B9,FtgDB,4,FALSE)**, from the footing database to calculate the width in inches, and nests the formula in the volume calculation formula.

The formula divides the width by 12 and the thickness by 12 to get the width and thickness in feet. The nested formula, **VLOOKUP(B9,FtgDB,4,FALSE)** returns the footing width (in inches) from the Footing Database and the nested formula, **VLOOKUP(B9,FtgDB,5,FALSE)**, is the footing thickness (in inches) retrieved from the Footing Database. The result is divided by 27 to get CY. A 5% waste factor is added and the amount is rounded up to the nearest CY.

The concrete unit costs and the labor unit costs ($/Unit) use VLOOKUP formulas to retrieve the correct unit costs from the footing database.

- Enter the following concrete unit cost formula into cell F9 of the *Footing* worksheet is as follows:

$$=VLOOKUP(B9,FtgDB,7,FALSE)$$

- Enter the following formula into cell H9 to calculate the labor cost per unit:

$$=VLOOKUP(B9,FtgDB,8,FALSE)$$

On the *Footing* worksheet, the concrete total costs (cells G9:G10) are the products of the CYs and the concrete unit cost. The labor total costs (cells I9:I10) are the products of the QTY (LF of footing) and the footing labor unit cost.

Set up the rebar and miscellaneous detail section similar to the footing concrete and labor details.

- Using **Data Validation** in cells B17:B18 on the *Footing* worksheet, select the appropriate description from the drop-down menu. (Be sure to name the description list in the rebar and miscellaneous database—cells B18:B25 on the *FootingDB* worksheet—so your formulas will use it as the source of the list.)

- Enter a VLOOKUP formula in cell C17 to return the quantities from the QTY column of the rebar and miscellaneous database. Formulas that calculate quantities are stored in the QTY column of the database (fig. 6.6).

Calculating the quantity of rebar for footings is easy. Typically, footings have two continuous horizontal #4 bars, although the size, grade, and number of bars vary from job to job. Rebar is available in 20′ lengths but the ends should be overlapped approximately 40 bar diameters—for #4 bar that means about 20″, or almost 2′. The practical length of a bar, then, is only about 18′.

Figure 6.6. Formulas view of Rebar and Miscellaneous database

	A	B	C	D	E
15					
16		Rebar & Misc DB			
17		PSI Concrete	QTY	Unit	Supplier 1
18		20'	=ROUNDUP(LFBar*Num_of_bars/18*(1+Waste),0)	EA	9.16
19		30" J-Bar	=ROUNDUP(LFBar/JBarSpacing*(1+Waste),0)	EA	1.89
20		48" J-Bar ⇩	=ROUNDUP(LFBar/JBarSpacing*(1+Waste),0)	EA	2.75
21		4'-6" J-Bar	=ROUNDUP(LFBar/JBarSpacing*(1+Waste),0)	EA	2.95
22		Step-up	1	EA	25
23		Blockout	1	EA	15
24		Bulkhead	1	EA	15
25					

- Enter the following formula in cell C18 of the rebar and miscellaneous database to calculate the correct quantity of bars:

=ROUNDUP(LFBar*Num_of_bars/18*(1+Waste),0)

In this equation

LFBar is the linear feet of footings.

Num_of_bars is the number of horizontal bars in the footings (1, 2, 3, etc.).

The divisor is the J-Bar spacing (the space between the vertical J-bars in inches, such as 18, 24, 30, etc.). **Waste** is the percentage of waste (.03, .05, 0.1, etc.).

Num_of_bars (C3), *LFBar* (C13), *JBarSpacing* (C4), and *Waste* (C5) are named cells on the footing take-off section of the *Footing* detail sheet (fig. 6.5, C3:C5). If you are creating your own worksheet, name these cells by selecting each cell and then typing the names in the name box, as previously discussed.

- Enter the following formula into cell C19 of the *FootingDB* worksheet to calculate the proper quantity of vertical J-bars:

=ROUNDUP(LFBar/JBarSpacing*(1+Waste),0)

To get the Unit and cost-per-unit ($/Unit), you must tell Excel® where to look up information in the rebar and miscellaneous database.

- Enter the following VLOOKUP formula in cell D17 of the *Footing* detail sheet:

=VLOOKUP(B17,FTGRebarMiscDB,3,FALSE)

- Enter the following formula in cell E17 of the *Footing* detail sheet:

=VLOOKUP(B17,FTGRebarMiscDB,4,FALSE)

The subtotal cost is the product of the quantity times the unit cost (QTY × $/Unit = Subtotal $). Therefore, enter the following formula in cell F17 of the *Footing* detail sheet:

=C17*E17

These examples illustrate the methods you can use to create detail sheets. You can add other formulas and additional functions to meet your specific needs.

Estimating Permit and Connection Fees

Many builders, and even some local officials, get frustrated calculating permit and connection/impact fees. As a builder, you may have to contend with multiple entities, each with its own method of calculating permit and plan check fees. Many municipalities have tried to standardize the process and usually base their calculations on guidelines from their region's major building code.

The value of the improvements to a property typically determines the permit fees. Each municipality determines its valuation rates, which are sometimes based on *Building Valuation Data*, national average building costs published semiannually by the International Code Council (ICC)[2]. Figure 6.7 shows a sample database with valuation rates for different municipalities as well as typical plan check fees and state tax rates.

Figure 6.7. Valuation and Plan Check database

	Code Database UBC - 1997					Plus an additional	For Each __ Above Base Fee
	For Property Improvement Valuation From:		To:	Base Fee	For the 1st		
4	$	1	$ 500	$ 23.50	$ 1	0	$ 1.00
5	$	501	$ 2,000	$ 23.50	$ 500	3.05	$ 100.00
6	$	2,001	$ 25,000	$ 69.25	$ 2,000	14.00	$ 1,000.00
7	$	25,001	$ 50,000	$ 391.25	$ 25,000	10.10	$ 1,000.00
8	$	50,001	$ 100,000	$ 643.75	$ 50,000	7.00	$ 1,000.00
9	$	100,001	$ 500,000	$ 993.75	$ 100,000	5.60	$ 1,000.00
10	$	500,001	$ 1,000,000	$3,233.75	$ 500,000	4.75	$ 1,000.00
11	$	1,000,001	$ 1,000,000,000	$5,608.75	$1,000,000	3.65	$ 1,000.00

The total building valuation is applied to a building permit fee table (fig. 6.8). Because the *International Building Code*®[3] did not include a permit table or a standardized plan check rate beginning with the 2000 code, many municipalities continue to use the permit table from the 1997 *Uniform Building Code*[4].

To determine the total valuation of improvements

1. Multiply the SF of the specific type of construction (e.g., main floor, garage, deck, etc.) by its valuation rate.

2. Calculate the permit fee using the graduated fee structure.

Suppose you are building a residence in Garden City that includes 1,950 SF on the main floor, 800 SF on the second floor, a 2,100 SF unfinished basement, a 650 SF garage, and a 150 SF covered porch (fig. 6.9). The total valuation would be $307,750.

■ Type the quantities—the floor SF of the types or locations of construction—into the QTY column of the *PermitConnectionFees* worksheet (C4:C11) or link to them if they are in another worksheet that stores key information.

Figure 6.8. Permit table

	Code Database UBC - 1997					
	For Property Improvement Valuation From:	**To:**	**Base Fee**	**For the 1st**	**Plus an additional**	**For Each __ Above Base Fee**
4	$ 1	$ 500	$ 23.50	$ 1	$ 0	$ 1.00
5	$ 501	$ 2,000	$ 23.50	$ 500	3.05	$ 100.00
6	$ 2,001	$ 25,000	$ 69.25	$ 2,000	14.00	$ 1,000.00
7	$ 25,001	$ 50,000	$ 391.25	$ 25,000	10.10	$ 1,000.00
8	$ 50,001	$ 100,000	$ 643.75	$ 50,000	7.00	$ 1,000.00
9	$ 100,001	$ 500,000	$ 993.75	$ 100,000	5.60	$ 1,000.00
10	$ 500,001	$ 1,000,000	$3,233.75	$ 500,000	4.75	$ 1,000.00
11	$ 1,000,001	$ 1,000,000,000	$5,608.75	$1,000,000	3.65	$ 1,000.00

Figure 6.9. Valuation calculations

		Municipality	Garden City	
	Price / Per SF	**Qty**	**Unit**	**Valuation Rate**	**Total Valuation**
4	Main Floor SF	1950	SF	$ 75.00	$ 146,250.00
5	2nd Floor SF	800	SF	$ 75.00	$ 60,000.00
6	3rd Floor SF		SF	$ 75.00	$ -
7	Finished Basement SF		SF	$ 55.00	$ -
8	Unfinished Basement SF	2100	SF	$ 35.00	$ 73,500.00
9	Garage SF	650	SF	$ 35.00	$ 22,750.00
10	Covered Porch SF	150	SF	$ 35.00	$ 5,250.00
11	Deck		SF	$ 35.00	$ -
13				Total Valuation	$ 307,750.00

You can use the floor SF stored in a key information sheet not only to determine valuations and fees, but to calculate framing materials and labor, finish carpentry materials and labor, drywall, painting, cleaning, allowances, and other items. Altering the SF in the key information sheet updates all of the detail sheets automatically.

You can also use links to connect information in your estimating worksheets, as discussed in Chapter 5. For example, if you named a cell *MainFloorSF* in your key information sheet, you could enter the following link in cell C4 of the *Permit-ConnectionFees* worksheet:

=MainFloorSF

This link copies to cell C4 whatever value is stored in the *MainFloorSF* cell.

Excel® will look up the valuation rate from the valuation database (cells B4:H12 of the *ValuationandFeeDB* worksheet) using the VLOOKUP formula and MATCH function.

■ Enter this formula in cell E4:

=VLOOKUP(B4,ValuationDB,MATCH(City,Cities,0)+1,FALSE)

The MATCH function returns the column number of Garden City (E2) from the list of cities to the column index argument of the VLOOKUP function. The +1 accounts for the one-column offset of the cities database from the valuation database (cells B4:H12, *ValuationandFeeDB*). You can use the Total Valuation (F13) on the *PermitConnectionFees* worksheet to calculate the permit fee.

Figure 6.10 shows the code database with formulas using the 1997 code that many localities still reference. Other codes may have similar fee structures. Once the building valuation is calculated, determine the permit fee by choosing the appropriate range in columns B and C of the *PermitCodeDB* worksheet. The valuation, $307,750, falls within the range of 100,001 and 500,000.

For the first $100,000 of valuation, the building permit fee is $993.75. For each additional $1,000 of valuation (rounded up) you must add $5.60. This is the mathematical process for calculating the fee:

1. Begin with $993.75.
2. Subtract $100,000 from $307,750, and round up the result to $208,000 (307,750 − $100,000 = $207,750).
3. Divide $208,000 by $1,000 to get 208.
4. Multiply 208 by $5.60, which equals $1,164.80.

The total permit fee is $2,158.55 ($993.75 + $1,164.80).

You can program Excel® to do the math for you as follows:

■ Enter the following formula in cell H9:

=IF(AND(TotalValuation>=B9,TotalValuation<=C9),

D9+ROUNDUP((TotalValuation-E9)/G9,0)*F9,“ ”)

This formula uses the IF function to determine which range contains the building's valuation. If the building valuation on the *PermitConnectionFees* worksheet

Figure 6.10. Permit table with formulas

| H9 | fx =IF(AND(TotalValuation>=B9,TotalValuation<=C9),D9+ROUNDUP((TotalValuation-E9)/G9,0)*F9,"") |

	A	B	C	D	E	F	G	H	I
1									
2		Code Database UBC - 1997							
3		For Property Improvement Valuation From:	To:	Base Fee	For the 1st	Plus an additional	For Each __ Above Base Fee		
4		$ 1	$ 500	$ 23.50	$ 1	0	$ 1.00		
5		$ 501	$ 2,000	$ 23.50	$ 500	3.05	$ 100.00		
6		$ 2,001	$ 25,000	$ 69.25	$ 2,000	14.00	$ 1,000.00		
7		$ 25,001	$ 50,000	$ 391.25	$ 25,000	10.10	$ 1,000.00		
8		$ 50,001	$ 100,000	$ 643.75	$ 50,000	7.00	$ 1,000.00		
9		$ 100,001	$ 500,000	$ 993.75	$ 100,000	5.60	$ 1,000.00	$2,158.55	
10		$ 500,001	$ 1,000,000	$3,233.75	$ 500,000	4.75	$ 1,000.00		
11		$ 1,000,001	$ 1,000,000,000	$5,608.75	$1,000,000	3.65	$ 1,000.00		
12							Permit Fee	$2,158.55	
13		Total Valuation	$ 307,750.00						

(F13) is greater than or equal to the value in column B and less than or equal to the value in column C, then the IF function will perform the following calculation:

D9+ROUNDUP((TotalValuation-E9)/G9,0)*F9

Otherwise, it will leave the cell blank ("").

- Copy the formula to cells H4 through H11 of the *PermitCodeDB* worksheet. The result of the calculation will only appear in cell H9 if the building valuation falls within the range (H4:H11) adjacent to the cell that contains the formula. In all the other cases the building valuation is outside the respective ranges (H4:H8 and H10:H11), so Excel® will leave the cell blank (""). The part of the formula that calculates the permit fee is:

D9+ROUNDUP((TotalValuation-E9)/G9,0)*F9

If you substitute values for the variables, the formula is:

=993.75+ROUNDUP(($307,750.00-$100,000)/$1,000,0)*5.6

Cell H12 is named *PermitFee*. The formula in cell H12 is the sum of cells H4:H11, (=SUM(H4:H11)). The Total Valuation shown in C13 is linked from cell F13 of the *PermitConnectionFees* worksheet. The Total Valuation is displayed in C13 of the *PermitCodeDB* to help you check the total.

Now, expand the *PermitConnectionFees* worksheet to include the state tax, permit, plan check, and connection fees (fig. 6.11).

- Enter the formula **=PermitFee** in cell E16.

The plan check fee is typically a percentage of the permit fee. Therefore, C17 also contains the formula **=PermitFee**.

Figure 6.11. Permit detail sheet

	A	B	C	D	E	F
1						
2				Municipality	Garden City	
3		Price / Per SF	Qty	Unit	Valuation Rate	Total Valuation
4		Main Floor SF	1950	SF	$ 75.00	$ 146,250.00
5		2nd Floor SF	800	SF	$ 75.00	$ 60,000.00
6		3rd Floor SF		SF	$ 75.00	$ -
7		Finished Basement SF		SF	$ 55.00	$ -
8		Unfinished Basement SF	2100	SF	$ 35.00	$ 73,500.00
9		Garage SF	650	SF	$ 35.00	$ 22,750.00
10		Covered Porch SF	150	SF	$ 35.00	$ 5,250.00
11		Deck		SF	$ 35.00	$ -
12						
13					Total Valuation	$ 307,750.00
14						
15		Permit and Connection Fees				
16		Permit Fee	1	Ea	$ 2,158.55	$ 2,158.55
17		Plan Check	$ 2,158.55	Valuation	$ 0.65	$ 1,403.06
18		State Tax	$ 2,158.55		$ 0.01	$ 21.59
19		Connection Fees	1	EA	$10,376.00	$ 10,376.00
20						
21					Total	$ 13,959.19

Cells E17 and E18 look up the Plan Check and State Tax rates from cells F16 and F17 in the *ValuationandFeeDB* worksheet using the VLOOKUP and MATCH functions.

■ Enter the following formula in cell E17:

=VLOOKUP(B17,PermitDataDB,MATCH(City,Cities,0),FALSE)

The formula looks up the Plan Check rate from the permit database on the *ValuationandFeeDB* worksheet (B16:H18).

■ Enter the following formula in cell E18:

=VLOOKUP(B18,PermitDataDB,MATCH(City,Cities,0),FALSE)

You can calculate connection fees one of two ways. Figure 6.12 shows a connection fees database, an additional database below the permit database.

One way to handle the connection fee is to look up the total connection fee for the city (Garden City, cell E19).

Figure 6.12. Connection fees database

Fee Data	Building Standards - 2008	Your City	Benton	Garden City	Lincoln	Windsor	
Phone							
Sewer Connection				984.00	50.00	225.00	
Sewer Impact			460.00		1,348.00		
Sewer District Capital Fac.			1,000.00			1,000.00	
Storm Water				712.00			
Pressurized Irrigation 1"			1,010.00				
Construction Water							
Water Connection			200.00	1,078.00	892.00	450.00	3/4"
Water Impact			940.00		200.00	765.00	
Meter Set Only							
Electrical Connection (Meter Hookup)			540.00	185.00			
Electrical Impact			233.00	641.00			
Parks & Recreation Impact			700.00	2,625.00	1,182.00	755.00	
Police & Safety						102.00	
Bonds (refundable)				3,000.00	1,000.00		
Temporary Power			80.00				
Construction Water Use				120.00	50.00	30.00	
Electrical Permit							
New service up to 200 amp			100.00				
Plumbing Permit							
# of fixtures					7.50	6.00	
Water heater and vent			20.00	17.00			
Gas line - up to 5 outlets							
Mechanical Permit							
New Furnace					22.50	18.50	
Appliance Vents			13.50				
Road Impact Fee			1,020.00	789.00			
Fire & EMS Sevices				225.00		24.00	
Miscellaneous							
Total Fees		0.00	6,316.50	10,376.00	4,752.00	3,375.50	
Water Connection Fees							
3/4"			200.00	1,078.00	892.00	450.00	
1"			310.00	1,781.00	1,070.00	500.00	
1-1/2"			1,500.00		2,450.00	1,775.00	
2"			3,500.00		4,500.00	3,200.00	

- Enter the following formula in cell E16:

=VLOOKUP("Total Fees",ConnectionDB,MATCH
(City,Cities,0)+1,FALSE)

VLOOKUP finds the row for Total Fees in the *ConnectionDB* (B22:H51 on the *ValuationandFeeDB* worksheet). The MATCH function returns the appropriate fee according to the city selected in cell E2 of the *PermitConnectionFees* worksheet.

You could also display more specifics about the connection fees on the detail sheet as follows:

- **Copy** the connection fee description list from the database.

- **Paste** it to the bottom of the detail sheet (fig. 6.13).

- Enter the following formula in cell C25 of the *PermitConnectionFees* detail sheet:

=VLOOKUP(B25,ConnectionDB,MATCH(City,Cities,0)+1,FALSE)

Figure 6.13. Connection fees detail

	A	B	C	D	E
22					
23		**Connection Fee Breakdown**	**Municipality**		
24		Phone	**Garden City**		
25		Sewer Connection	$ 984.00		
26		Sewer Impact	$ -		
27		Sewer District Capital Fac.	$ -		
28		Storm Water	$ 712.00		
29		Pressurized Irrigation 1"	$ -		
30		Construction Water	$ -		
31		Water Connection	$ 1,078.00	Water Line	3/4"
32		Water Impact	$ -		To Database
33		Meter Set Only	$ -		
34		Electrical Connection (Meter Hookup)	$ 185.00		
35		Electrical Impact	$ 641.00		
36		Parks & Recreation Impact	$ 2,625.00		
37		Police & Safety	$ -		
38		Bonds (refundable)	$ 3,000.00		
39		Temporary Power	$ -		
40		Construction Water Use	$ 120.00		
41		Electrical Permit	$ -		
42		New service up to 200 amp	$ -		
43		Plumbing Permit	$ -		
44		# of fixtures	$ -		
45		Water heater and vent	$ 17.00		
46		Gas line - up to 5 outlets	$ -		
47		Mechanical Permit	$ -		
48		New Furnace	$ -		
49		Appliance Vents	$ -		
50		Road Impact Fee	$ 789.00		
51		Fire & EMS Sevices	$ 225.00		
52		Miscellaneous	$ -		
53					
54		Total Fees	$ 10,376.00		
55					

- **Copy** the formula down to cells C26:C53.
- Enter the following Total Fees formula in cell C54:

=SUM(C25:C52)

Cell E31 of the *PermitConnectionFees* detail sheet offers a choice of water line sizes. When selected, the Water Connection fee in the *ConnectionDB* should reflect the proper fee for the selected size of water line. Enter the following formula into cell E28 of the *ConnectionDB* on the *ValuationandFeeDB* worksheet. The formula will adjust the fee according to the water line size selected:

=VLOOKUP(WaterLineSize,WaterConnectionDB,4,FALSE)

- Enter similar formulas in cells F28:H28 of the *ValuationFeeDB*, substituting 5, 6, or 7 for the column index number.

In figure 6.12, cell I28 of the *ValuationandFeeDB*, the selected line size is displayed so you can quickly verify that it is correct.

Working with Variables

As discussed previously, you should not include variables (e.g., the SF of a main floor, the valuation of a home, or the quantity of horizontal rebar in a footing) in formulas and you should not use cell addresses in formulas. If the variables change, or if you add rows or columns to your worksheets, you will have to rewrite the formulas. Instead, name cells and cell ranges and use these names in your formulas.

Adding Comments to Cells

Occasionally, you may wish to clarify or provide special instructions about information in a specific cell by inserting a comment (fig. 6.14). You can insert comments into cells and they will appear when you hover the cursor over the cell.

Figure 6.14. Adding comments to cells

	A	B	C	D	E	F	G	
19								
20		Fee Data	Building Standards - 2008	Your City	Benton	Garden City	Lincoln	Wi
21		Phone						
22		Sewer Connection				984.00	50.00	
23		Sewer Impact			460.00		1,348.00	
24		Sewer District Capital Fac.			1,000.00			1,
25		Storm Water				712.00		
26		Pressurized Irrigation 1"			1,010.00			
27		Construction Water						
28		Water Connection			200.00	1,078.00	892.00	
29		Water Impact			940.00		200.00	
30		Meter Set Only						
31		Electrical Connection (Meter Hookup)			540.00	185.00		
32		Electrical Impact			233.00			
33		Parks & Recreation Impact			700.00	Meadow Brook subdivision is exempt from this fee		
34		Police & Safety						
35		Bonds (refundable)						
36		Temporary Power			80.00			
37		Construction Water Use				120.00	50.00	
38		Electrical Permit						

To insert a comment into a cell or range of cells

- Select the cells where you wish to add a comment.
- Either right click and select **Insert Comment** from the pop-up menu or click **New Comment** on the **Review** tab of the ribbon.

A comment box will appear next to the cell. The comment box is in edit mode and is ready for you to type in notes or instructions. You can edit or delete comments as follows:

- Right click the cell.
- Click **Edit Comment** or **Delete Comment** from the pop-up menu.
- If you want a comment to remain visible when the cursor is not over the cell that contains the comment, right click the cell and then click **Show/Hide Comment**.
- To hide the comment so it appears only when the cursor passes over the cell, right click the cell and click **Hide Comment**.

Calculating Profit Margin

Builders and remodelers try to make their estimates accurate to within 1%–2% of their actual costs. Yet many builders and remodelers lose more money by miscalculating their profit and overhead than they do by calculating quantities erroneously. In fact, many of these builders make only about 66% of the profit they thought they would because they estimate their profit margins incorrectly. Miscalculating margin, which includes both company overhead and profit, is a costly mistake. This chapter explains how to use Excel® to calculate construction loan interest and builder's margin so you will know exactly what your real profit is.

Construction Loans

Construction loan costs typically include interest, title insurance, and the following fees:

- origination
- appraisal
- inspection
- closing
- miscellaneous

Whether or not you use a construction loan, the time and money you spend during the construction phase are opportunity costs—they have value that could be invested elsewhere. If you do not enter an amount in the estimate to account for these costs, you erode the final profit margin.

Figure 7.1 shows a construction loan detail sheet. Named ranges on the sheet (in parentheses) are as follows: sales price or appraised value (*AppraisedValue*), loan-to-value ratio (LTV), origination fee rate (*OriginationFeeRate*), annual percentage rate (APR), and the term of the loan (*ConstLoanTerm*). The loan amount

Figure 7.1. Construction loan detail

	B	C	D	E	F
2	**Construction Loan Detail**				
3					
4	Appraised Value: $	350,000		Origination Rate:	1.00%
5	LTV: $	0.80		Interest Rate APR:	6.80%
6	Loan Amount: $	280,000		Term: (Months)	6
7					
8					
9	**Construction Loan**				Lender 1
10	Description	QTY	Unit	$/Unit	Sub-Total $
11	Loan Origination Fee	$ 280,000	EA	1.00%	$ 2,800.00
12	Interest Reserve	$ 280,000	EA	0.017	$ 4,760.00
13	Document Preparation	1	EA	$ 50.00	$ 50.00
14	Title Insurance	1	EA	$ 1,800.00	$ 1,800.00
15	Closing Fees	1	EA	$ 500.00	$ 500.00
16	Flood Certificate	1	EA	$ 17.50	$ 17.50
17	Hazard Insurance	1	EA	$ 350.00	$ 350.00
18					
19					
20				Total	$ 10,277.50
21					

($ConstLoanAmount$) is calculated as a percentage (LTV) of the appraised value. The loan term is typically expressed in months.

The loan origination fee and interest reserve (the amount set aside to cover interest charges for money borrowed during the course of construction) are expenses typical to all construction loans. Other fees may be typical but their amounts vary among lenders.

Estimating the Interest Reserve

You can estimate the amount of the interest reserve using the cumulative costs of construction. Figure 7.2 shows a typical construction S curve. It was generated using the monthly accrued job costs displayed in table 7.1.

Costs accrue slowly at the beginning of construction, but as a project progresses and more trade contractors complete their work, costs increase rapidly. As the project winds down to the finishing stage, you hire fewer trade contractors and make fewer materials purchases, so costs level off. Figure 7.3 shows the S curve

TABLE 7.1. Monthly and accrued job costs

Month	Monthly Costs	Accrued Job Costs
Month 0	$ —	$ —
Month 1	$ 21,140	$ 21,140
Month 2	$ 49,000	$ 70,140
Month 3	$ 69,020	$ 139,160
Month 4	$ 72,100	$ 211,260
Month 5	$ 45,640	$ 256,900
Month 6	$ 23,100	$ 280,000

Figure 7.2. Construction S curve

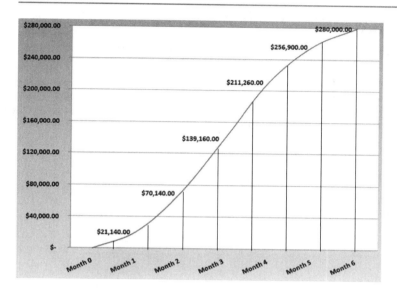

Figure 7.3. Making a rectangle using the S curve

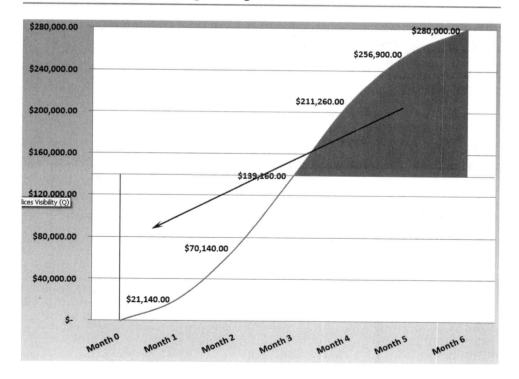

with the area under the line representing the total accrued costs over the construction period.

The monthly interest rate multiplied by the area under the curve—the total accrued costs of construction—is the interest expense for the project. If you knew the formula for the construction curve line, you could use calculus to find the area of the space under the line. But you really don't need calculus. Instead, you can estimate the size of the area under the curve using a simpler technique. Just flip and move the shaded area that is under the upper part of the curve to the lower part of the graph that is above the curve (fig. 7.4) to make a rectangle. It is much easier to estimate the area of this rectangle than the other two shapes. This rectangle represents the total accrued costs of construction.

The interest formula, which multiplies the monthly interest by the approximate area under the curve, is as follows:

$$= \text{Monthly Interest Rate} * 1/2 \text{ Loan Amount} * \text{Term in Months}$$
$$= \text{APR}/12 * .5 * \text{ConstLoanAmount} * \text{ConstLoanTerm}$$

Although lending institutions use different formulas—some lenders will use a 0.6 factor instead of 0.5, setting aside more interest reserve money—the 0.5 factor is usually adequate. It may be more than enough if the builder pays final expenses of the project from long-term financing instead of a construction loan draw.

Figure 7.4. Using the rectangle to estimate construction costs

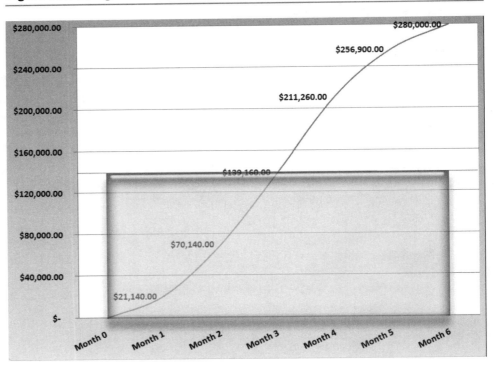

Builder's Margin: Profit and Company Overhead

Builders must add profit and company overhead to their direct construction costs to determine a sales price. Unfortunately, many builders incorrectly calculate profit. NAHB's biennial *The Cost of Doing Business Study*[5] has found that builders routinely overestimate what their profit will be. The problem is not inefficiency or ineffectiveness, but rather the method they use to calculate profit.

A company's annual gross sales can be divided into direct costs, company overhead, and profit (fig. 7.5).

Direct costs are hard costs, such as materials, labor, and equipment; and project overhead, such as permits, fees, and temporary utilities. Any cost that you can attribute to a specific job is a direct cost.

Company overhead, often referred to as general and administrative (G&A) costs, comprises the costs of doing business, including officer salaries, advertising, liability insurance, and main office expenses.

Profit or net profit is the reward for the risk that builders and remodelers take to complete a project. It is the money that remains after bills are paid. Gross profit includes both company overhead and net profit.

Profit and company overhead often are expressed as percentages of gross sales. Suppose your company overhead consumed 8% and profit was 10% of sales. Your sales for the coming year are expected to be comparable to this year's so the company overhead is not expected to change.

Your estimated costs for an upcoming job are $200,000. To determine the profit and company overhead to include in a bid, you want to incorporate your customary company overhead (8%) and profit (10%). What dollar amount should you include for profit and company overhead in the sales price? Take a minute to write your answer. Do not consider or include the cost of the lot yet; you will add the lot cost after the profit and overhead are added to the building costs.

Figure 7.5. Breakdown of project costs

Sales Price (Gross Income)

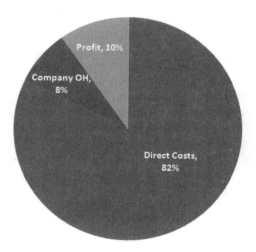

Did you write $236,000 ($200,000 × 1.18), or even $237,600 ($200,000 × 1.08 × 1.10)? If you used either of these calculations, your profit will be much less than what you expect. Figure 7.6 shows that direct costs are 82% of the total sales price. To find the total sales price, divide both sides of the equation by .82 as follows:

$$\frac{Direct\ Costs}{.82} = \frac{.82 \times total\ Sales\ Price}{.82}$$

$$Total\ Sales\ Price = \frac{Direct\ Costs}{.82}$$

When you apply the formula, you get a sales price of $243,902.44. To find the company overhead, multiply the sales price by 8%. To determine the profit, multiply the sales price by 10%.

$$Company\ Overhead = \$243,904.44 \times .08 = \$19,512.20$$
$$Profit\ Margin = \$243,902.44 \times .10 = \$24,390.24$$

Suppose you had determined the sales price by multiplying direct costs of $200,000 by 1 plus the margin to get $236,000 ($200,000 × 1.18). To determine the percentage of profit you will earn, subtract the direct costs and company overhead from the sales price. Whether or not the correct overhead was estimated, it has to be paid, leaving the surplus as profit.

Sales Price	$236,000.00
Direct Costs	− $200,000.00
	$36,000.00
Overhead	− $19,512.20
Profit	$16,487.80

Profit Percentage = 0.0699 (less than 7%)

Figure 7.6. Breakdown of project costs–direct cost formula

Sales Price (Gross Income)

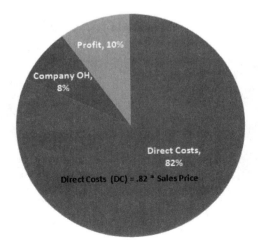

Remember, the target profit was 10%. In order to achieve it, the sales price must account for the profit in addition to direct costs and overhead. Do not make the mistake of calculating profit only as a percentage of direct costs. Profit margin is profit as a percentage of the whole pie (gross sales). This method is used by all retail businesses, by most accountants, and by all major building companies. Profit markup, on the other hand, is only a percentage of direct costs. This method is used in most commercial, industrial, heavy, and highway construction.

To correctly calculate a sales price, use a multiplier and the direct costs of a construction project. To find a multiplier that correctly calculates the sales price as the product of the direct costs and the multiplier, divide the direct costs (using a unit of 1 as the direct cost) by 1 minus the margin as follows:

$$\text{Multiplier} = \text{Direct Costs} \div (1 - \text{Margin}) = 1 \div (1 - \text{Margin})$$

A margin of 10% uses a multiplier of 1.11:

$$\text{Multiplier} = 1 \div (1 - .10) = 1 \div .90 = 1.11$$

A margin of 15% uses a multiplier of 1.18.
A margin of 20% uses a multiplier of 1.25.

Table 7.2 lists other markups and margins, and their corresponding multipliers. Figure 7.7 shows the spreadsheet Margin v. Markup with the formulas view on so you can see the formulas used to create the table.

Remember, you can toggle on and off the formulas view by pressing **Ctrl +** ` (accent key).

Figure 7.7. Formulas to compute margin and markup

	A	B	C	D	E	F
1						
2		Margin	Markup		Markup	Margin
3		1	=1/(1-(B3/100))		1.01	=-100*((1/E3)-1)
4		2	=1/(1-(B4/100))		1.02	=-100*((1/E4)-1)
5		3	=1/(1-(B5/100))		1.03	=-100*((1/E5)-1)

Margin vs. Markup

Builders and remodelers may define profit either as markup or margin, which can cause problems with customers, especially in a cost-plus contract, i.e., a contract in which you agree to do a project for "cost" plus a specified additional percentage. The best way to avoid misunderstandings in these contracts is to specify whether you are using a markup or builder's margin.

Although you may prefer to calculate profit by applying a markup rather than a margin because you think that is what your competitors are doing, consider that using markups will not make you more competitive. If you have to compete by lowering your profit, do so by using a lower percentage rate (7% instead of 10%, for example) and calculate the margin, which represents your true profit percentage. Above all, do not try to fool yourself by calling your 10% markup a 10% profit rather than the 7% profit that it is. And, maintain clarity about your bottom line by referring to builder's margin rather than profit.

Figure 7.8 is taken from the *CalculatingProfit* worksheet in the Chapter 7 workbook. It shows how direct costs, company overhead, and margin are totaled to arrive at the sales price.

TABLE 7.2. Markups, margins, and multipliers

Margin	Markup	Markup	Margin	
1	1.010101	1.01	0.990099	Example: To get a
2	1.020408	1.02	1.960784	15% margin, use a 1.176471 markup
3	1.030928	1.03	2.912621	
4	1.041667	1.04	3.846154	Example: To get a
5	1.052632	1.05	4.761905	15% markup, use a 13.0438% margin
6	1.06383	1.06	5.660377	
7	1.075269	1.07	6.542056	
8	1.086957	1.08	7.407407	
9	1.098901	1.09	8.256881	
10	1.111111	1.1	9.090909	
11	1.123596	1.11	9.90991	
12	1.136364	1.12	10.71429	
13	1.149425	1.13	11.50442	
14	1.162791	1.14	12.2807	
15	1.176471	1.15	13.04348	
16	1.190476	1.16	13.7931	
17	1.204819	1.17	14.52991	
18	1.219512	1.18	15.25424	
19	1.234568	1.19	15.96639	
20	1.25	1.2	16.66667	
21	1.265823	1.21	17.35537	
22	1.282051	1.22	18.03279	
23	1.298701	1.23	18.69919	
24	1.315789	1.24	19.35484	
25	1.333333	1.25	20	
26	1.351351	1.26	20.63492	
27	1.369863	1.27	21.25984	
28	1.388889	1.28	21.875	
29	1.408451	1.29	22.48062	
30	1.428571	1.3	23.07692	
31	1.449275	1.31	23.66412	
32	1.470588	1.32	24.24242	
33	1.492537	1.33	24.81203	
34	1.515152	1.34	25.37313	
35	1.538462	1.35	25.92593	
36	1.5625	1.36	26.47059	
37	1.587302	1.37	27.0073	
38	1.612903	1.38	27.53623	
39	1.639344	1.39	28.05755	
40	1.666667	1.4	28.57143	
41	1.694915	1.41	29.07801	
42	1.724138	1.42	29.57746	
43	1.754386	1.43	30.06993	
44	1.785714	1.44	30.55556	
45	1.818182	1.45	31.03448	
46	1.851852	1.46	31.50685	
47	1.886792	1.47	31.97279	
48	1.923077	1.48	32.43243	
49	1.960784	1.49	32.88591	
50	2	1.5	33.33333	

Figure 7.8. Sales price computation

	A	B	C
34			
35		Total Direct Costs	$200,000.00
36			
37	0.08	Company Overhead	$ 19,512.20
38	0.1	Builder's Margin	$ 24,390.24
39			
40		Sales Price	$243,902.44
41			

Figure 7.9. Formulas used to determine sales price

	A	B	C
34			
35		Total Direct Costs	=SUM(C11,C13:C34)
36			
37	0.08	Company Overhead	=A37*SalesPrice
38	0.1	Builder's Margin	=A38*SalesPrice
39			⊕
40		Sales Price	=DirectCosts/(1-(A37+A38))
41			

The percentages used for the company overhead and builder's margin (cells A37 and A38) can be named and even referred to from a different worksheet. Figure 7.9 shows the formulas used.

This method of calculating the sales price is probably the most straightforward and is less prone to error than other methods. Another method uses Microsoft Excel®'s ability to calculate an amount through iteration by using circular references.

Circular References

If one cell's value depends on another's and vice versa, the result is a circular reference. A circular reference made in error is problematic. When Excel®'s "Iteration" is turned off (the default setting), the program will notify you that it cannot calculate a circular reference. However, with Iteration turned on, you can use a circular reference to determine a sales price (fig. 7.10).

Figure 7.10. Using iteration to determine sales price

	A	B	C
34			
35		Total Direct Costs	=SUM(C11,C13:C34)
36			
37	0.08	Company Overhead	=A37*SalesPrice
38	0.1	Builder's Margin	=A38*SalesPrice
39			
40		Sales Price	=SUM(C35:C38)
41			

The amount of company overhead and profit depend on the sales price. Conversely, the sales price depends on the company overhead and builder's margin. Excel® will solve an equation with circular references by plugging values into the variables and checking the results until the equation works.

To enable iteration

- Click the **Office button** and **Excel® Options** (at the bottom of the menu displayed) to reveal the **Excel® Options** pop-up box (fig. 7.11).
- Select **Formulas** on the left menu bar.
- Click the **Enable iterative calculation** box.

Iteration will be active in all workbooks until you uncheck the box. If you accidentally create a circular reference, you can use the **Trace Precedents** and **Trace Dependents** tools on the Formulas tab to locate the mistake (fig. 7.12).

The trace arrows point to cells with predecessor or dependent references. Circular references are powerful, time-saving tools.

Figure 7.11. Enabling iteration

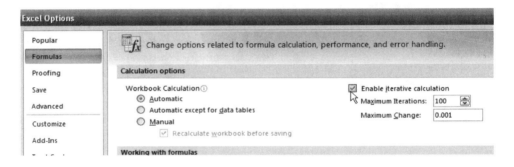

Figure 7.12. Tracing tools

	A	B	C
34			
35		**Total Direct Costs**	=SUM(C11,C13:C34)
36			
37	0.08	Company Overhead	=A37*SalesPrice
38	0.1	Builder's Margin	=A38*SalesPrice
39			
40		**Sales Price**	=SUM(C35:C38)

Making Spreadsheets User Friendly with Form Controls

8

Form controls are fun to work with, practical, and visually appealing. They will enhance the functionality of your estimating spreadsheets. To access Form controls you must first display the Developer tab on the ribbon.

- Click the **Office button** and **Excel® Options** (fig. 8.1).
- Click **Popular** on the left menu bar.
- Check the **Show Developer** tab in the ribbon box.

The Developer tab is now on the ribbon (fig. 8.2).

Figure 8.1. Show Developer tab

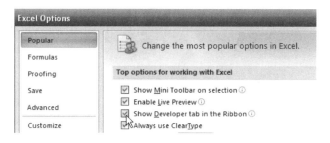

Figure 8.2. Controls

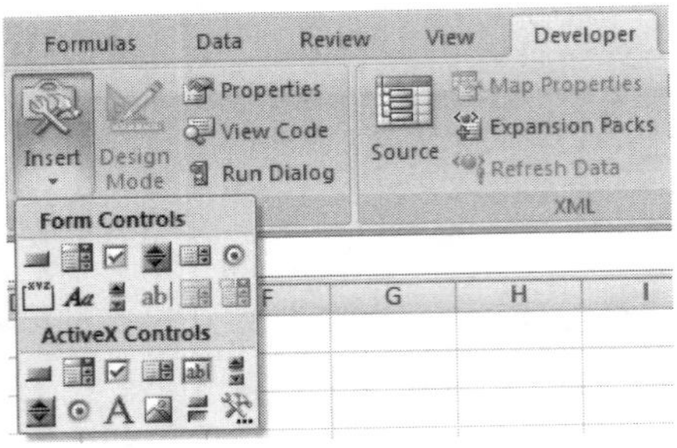

Figure 8.3. Using check boxes

A	B	C	D	E	F	G	H	I	J	K	L
1											
2	**Windows**										
3	QTY	Size	Unit Cost	Low E	9%	Tinted	12%	Grids	$ 12.50	Unit Total	Total Cost
4	2	4040	$ 163.00	☑ Low E	$ 14.67	☑ Tinted	$ 19.56	☐ Grids		$ 197.23	$ 394.46
5	3	5040	$ 184.00	☑ Low E	$ 16.56	☑ Tinted	$ 22.08	☐ Grids		$ 222.64	$ 667.92
6	1	5050	$ 225.00	☐ Low E		☐ Tinted		☑ Grids	$ 12.50	$ 237.50	$ 237.50
7											
8										Tax	$ 84.49
9										Total	$ 1,384.37

Check Boxes

The **Check Box** control allows a user to decide between two options by clicking on the box. You can use check boxes to answer yes-or-no questions about an item. When checked, the value of box is "true" and when unchecked, its value is "false." You can store the value ("true" or "false") in a linked cell. You must link each check box to a different cell. For example, in choosing a window you might want to select features such as low E glass, tinted glass, or grids.

Windows are listed with their base unit costs (fig. 8.3 and *Checkboxes* worksheet in the Chapter 8 workbook). Excel® calculates additional costs for low E glass, tinted glass, and grids when any of these boxes are checked. The equations to calculate the window order are

$$\text{Unit Cost} + \text{Each Additional Cost} = \text{Unit Total}$$

and

$$\text{QTY} \times \text{Unit Total} = \text{Total Cost}$$

The Excel® formula for the Unit Total (K4) is

$$=\text{SUM(D4,F4,H4,J4)}$$

The Total Cost (Quantity × Unit Cost) is

$$=\text{B4*K4.}$$

Use the example in figure 8.3 to begin setting up the check boxes.

- Click **Insert** on the **Developer** tab.
- Choose **Check Box** from the options displayed on the pop-up menu.

- Choose where you want to place the check box in your worksheet. (Place the check box in cell E4.)
- To label the check box "Low E," right click it and choose **Edit Text** from the pop-up menu.

You can reposition the check box using one of the following two methods:

1. Right click the check box, press the **Esc** key, and use the **arrow keys** on your keyboard to move the check box.
2. Click the border of the check box frame and drag it to the new location. After you create and label the first check box correctly, you can copy it to create check boxes for other items.

The next step is to create a link for each check box. You can link the value of the check box (again, "true" if checked and "false" if unchecked) to another cell. To create the link

- Right click the check box and choose **Format Control** from the pop-up menu.
- Click the **Control** tab on the **Format Control** box that pops up, and click the **Cell link** edit box.
- Type the cell address into the edit box or click the button to the right of the edit box. (The **Format Control** box will shrink to show only the **Edit** box.)
- Click the cell that you want to link to. When the box is checked the value of E3 is true, and when it is unchecked the value is false.
- Press **Enter**.
- Click **OK** to finalize the cell link.

To hide the words "True" or "False" in E3 so the text doesn't overlap and clutter the check box, change the text color to white. The value of the cell will then be visible only in the formula bar.

- Enter a formula in cell F4 to increase the base cost of the window (D4) by the percentage increase for low E (F3).

For example, if adding low E coating to the windows increases the base cost by 9% (the value stored in cell F3), you would enter the following formula into cell F4 to correctly calculate the cost increase:

$$=IF(E4,D4*F3,"")$$

The formula reads, "If cell E4 is true (the Low E box is checked), then multiply the base unit cost (D4) by 0.09 (F3). If the value is false leave the value blank (""). The equation, $=IF(E4,D4*F3,"")$ is equivalent to $=IF(E4=True,D4*F3,"")$.
A better equivalent formula is

$$=IF(E4,D4*\$F\$3,"")$$

The $ signs to the left of the F and the 3 anchor these references so as you copy the formula down to other cells (F3 and F4), the reference to F2 does not change like the other relative references do (fig. 8.4). To anchor a cell reference

- Place the cursor in or next to the reference.
- Press the **F4 key**.

Figure 8.4. Anchoring a cell in a formula

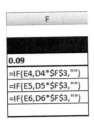

The F4 key will toggle through three options. In this example, the reference changes from F3 to F3 to F$3 to $F3. Pressing the key a fourth time would start the cycle again at F3. The dollar sign anchors whatever reference follows it.

Option Buttons

An option button allows users to select items from a list. Suppose that you had the following three options for the material for a supply water line:

- copper
- cross-linked polyethylene (PEX)
- chlorinated polyvinyl chloride (CPVC)

Using Option buttons, you can automatically update the unit cost based on the material. Figure 8.5 shows Option buttons for the types of pipe material, a detail of a supply water line take-off, and a piping material database. (The figure shows the database and detail sheet on one worksheet, but the database typically would be on a separate sheet.)

To create an Option button

- Click the **Option button** on the **Form controls** menu. The mouse pointer changes to a crosshair.

Figure 8.5. Selecting materials using Option buttons

Description	Qty	Unit	Unit Cost	Total
Supply Water Line	50	LF	$ 0.58	$ 29.00

Option buttons:
- ○ Copper
- ◉ PEX
- ○ CPVC

Pipe Database			
Description	Unit	Unit Cost	
1/2" Copper Type N	LF	$ 1.35	
1/2" PEX Clear	LF	$ 0.58	
1/2" CPVC	LF	$ 0.87	

■ Click the worksheet where you want to place the Option button and drag diagonally down and to the right until the button is the desired size.

■ Release the mouse button.

Repeat these steps for the other two Option buttons or copy the Option button as follows:

■ Right click with the mouse pointer over the **Option button**.

■ Clear the menu by pressing the **Esc** key.

■ While pressing Ctrl, click and drag the border of the Option button to the desired location.

■ Edit the text description.

To move or resize the Option button

■ Place the mouse pointer over it and right click.

■ Press **Esc** to clear the menu.

■ Click the edge of the border and drag the button to the desired location.

■ To resize the button, click one of its handles and drag to the desired size.

For Option buttons, the cell link is the cell that stores the value of the control (A2). To set the cell link

■ Right click any of the Option buttons (fig. 8.6).

■ Click the **Control** tab, and enter the cell address into the Cell link box. (You could instead click the cell where you want to place the cell link and Excel® will enter the address into the Cell link box.)

■ Click **OK**.

Cell A2 now contains the value of the Option button controls (1, 2, or 3).

■ Enter the following formula into cell E7 (fig. 8.7)

$$=IF(A2=1,D12,IF(A2=2,D13,D14))$$

The formula uses the information stored in cell A2 (the cell link for the Option buttons). Again, you can hide the text in cell A2 by making it white. The text will be invisible, even though the value of the cell will not change.

The syntax for the IF function is **=IF(logical_test, value_if_true, [value_if_false])**. Therefore, the formula in E7 reads as follows:

Figure 8.6. Control cell link used for Option button

Figure 8.7. Formula used with Option button

	E7	▾	*fx*	=IF(A2=1,D12,IF(A2=2,D13,D14))	
	A	B	C	D	E	F
1						
2	2	○ Copper				
3		⦿ PEX				
4		○ CPVC				
5						
6		**Description**	**Qty**	**Unit**	**Unit Cost**	**Total**
7		Supply Water Line	50	LF	$ 0.58	$ 29.00
8						
9						
10		**Pipe Database**				
11		**Description**	**Unit**	**Unit Cost**		
12		1/2" Copper Type N	LF	$ 1.35		
13		1/2" PEX Clear	LF	$ 0.58		
14		1/2" CPVC	LF	$ 0.87		
15						

- If the value in cell A2 equals 1, then the value in E7 is whatever value is stored in D12 ($1.35).
- If the value in cell A2 equals 2, then the value in E7 is whatever value is stored in D13 ($0.58).
- The only other option is 3, or whatever value is stored in cell D14 ($0.87).

If you want two or more sets of Option buttons on a worksheet, you must group each set so it will act independently.

- Select the **Group Box** button (the square icon with the XYZ) on the Form controls menu.
- Click and drag on the worksheet to draw a group box around each set of Option buttons (fig. 8.8).

The Option buttons in each group will link to the cell designated for that group and the Option buttons for other groups will link to their designated cells.

Scroll Bar

Scroll bars can help users input numbers or quantities (fig. 8.9). They are fun to create and add functionality to a worksheet. They work like the scroll bars in other Microsoft® programs. When a user clicks on the arrow, the value of the linked cell

Figure 8.8. Grouping Option buttons

	A	B	C	D	E
1		┌ Supply Pipe ───		┌ Sewer Pipe ───	
2	2	○ Copper	1	⦿ ABS	
3		⦿ PEX		○ PVC	
4		○ CPVC			
5					

Figure 8.9. Scroll bar

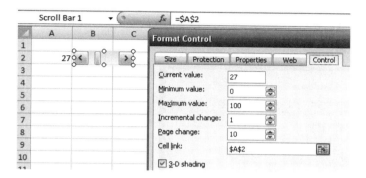

changes by one. Clicking between the arrows, on the bar, makes the change greater. You also can drag the scroll bar to the desired value. To show the value of the scroll bar, you must link it to a cell.

To create a scroll bar

- Click the **scroll bar icon** on the **Form Controls** menu.
- Click and drag in the area on the worksheet where you want to place the scroll bar. You can position the scroll bar horizontally or vertically.
- Right click the scroll bar and click **Format Control**. Be sure the **Control** tab is selected.
- Link the scroll bar to A2 by clicking on the button to the right of the **Cell link** edit box.
- Click on the desired cell (A2).
- While the Format Control screen is still visible, set the minimum, maximum, incremental, and page change values to best fit your application.

Spinner

The spinner works like a scroll bar except it will only change values incrementally (fig. 8.10). Suppose that you want to use a spinner to select the slope for calculating the squares of roof shingles.

- Name cell B2, *Slope* (Click cell B2, type *Slope* in the Name box, and press **Enter**). After you create the spinner, you will link it to this cell.
- To create the spinner, click the **spinner icon** on the **Form Controls** menu.
- Click and drag over the area on the worksheet where you want to place the spinner.
- Right click the spinner to format the controls for it. You must set the minimum value (minimum slope you want to allow), the maximum value (maximum slope you want to allow), and the incremental change.
- Type the named range *Slope* in the **Cell link** box and click **OK** at the bottom of the Format Control menu to specify the cell to be linked to the spinner.

Figure 8.10. Spinner

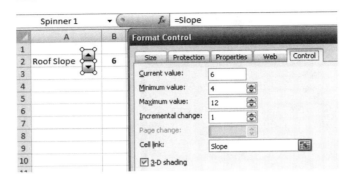

Cell B2 (Slope) now displays whatever value the spinner specifies.

List Boxes

List Boxes allow users to select items from a list. A List Box displays the items assigned to that list (fig. 8.11).

Creating a List Box control allows a user to click an item in the list, which Excel® will then place in a designated cell (fig. 8.12). For example, if you have a list of kitchen ranges customers can choose from, and you want to place this list,

Figure 8.11. List Box

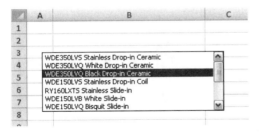

Figure 8.12. Making a List Box functional

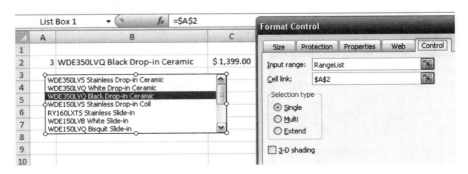

named *RangeList* (cells E11–E23 of the *ListBox* sheet) in a List Box, follow these steps:

- Select the **List Box** button from the **Forms control** menu and click and drag on the worksheet where you want to place the List Box.
- Right click the **List Box** to bring up the Format Control menu.
- Enter the named range (*RangeList*) into the **Input range** box and type the cell link (A2) into the **Cell link input,** or click cell A2 in the box.

When a user selects an item, Excel® stores its value, or its position in the list (3) in the linked cell (A2). To make the worksheet more user friendly, however, you want to display the name of the item selected.

The value of the position of the selected item is the value of the List Box. The position of "WDE350LVQ Black Drop-in Ceramic" in the range list is 3.

- Enter the following *index* formula in cell B2 to convert the position number to the name of the item (fig. 8.13):

$$= INDEX(RangeList,A2)$$

In this example, "WDE350LVQ Black Drop-in Ceramic" will appear in cell B2. As noted previously, you can format the text in cell A2 to be white so it will not show.

- Enter the following VLOOKUP formula in cell C2 to find the cost of the selected kitchen range in the *Range* database (*RangeDB*):

$$=VLOOKUP(B2,RangeDB,2,FALSE)$$

Figure 8.13. Formulas used with a List Box

	A	B	C
1			
2	3	=INDEX(RangeList,A2)	=VLOOKUP(B2,RangeDB,2,FALSE)
3		WDE350LVS Stainless Drop-in Ceramic	
4		WDE350LVQ White Drop-in Ceramic	
5		WDE350LVQ Black Drop-in Ceramic	
6		WDE150LVS Stainless Drop-in Coil	
7		RY160LXTS Stainless Slide-in	
8		WDE150LVB White Slide-in	
		WDE150LVQ Bisquit Slide-in	

Figure 8.14. Completed List Box

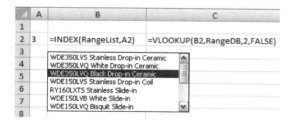

A2		f_x	3

	A	B	C
1			
2		WDE350LVQ Black Drop-in Ceramic	$ 1,399.00
3		WDE350LVS Stainless Drop-in Ceramic	
4		WDE350LVQ White Drop-in Ceramic	
5		WDE350LVQ Black Drop-in Ceramic	
6		WDE150LVS Stainless Drop-in Coil	
7		RY160LXTS Stainless Slide-in	
8		WDE150LVB White Slide-in	
		WDE150LVQ Bisquit Slide-in	

Figure 8.14 shows the completed List Box. You also could use VLOOKUP to look up an item's unit of measure (EA, LF, SF, etc.) in a database.

Combo Boxes

A Combo Box is similar to a List Box; however, it displays only the selected item until a user clicks the down arrow (fig. 8.15). You create it the same way you create a List Box.

Make the cell link and write the formulas the same way you did for the List Box. Figure 8.16 shows the appliance selection from the list.

Figure 8.15. Combo Box

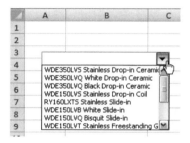

Figure 8.16. Creating a Combo Box

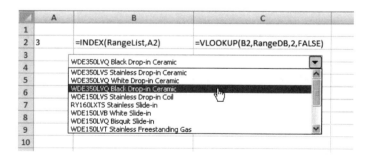

Figure 8.17. Completed Combo Box

Figure 8.18. Making a selection with a Combo Box

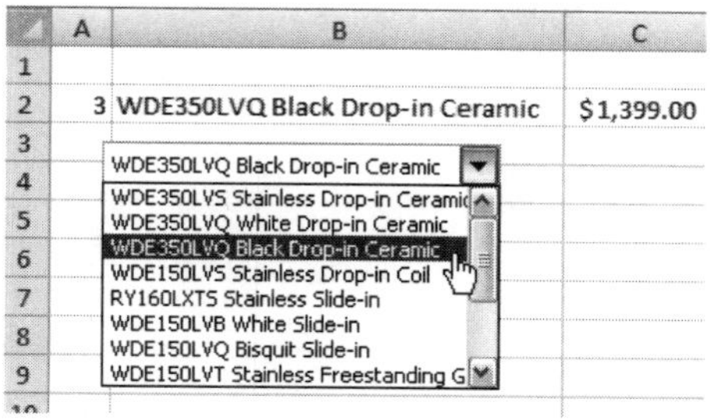

	A	B	C
1			
2	3	WDE350LVQ Black Drop-in Ceramic	$1,399.00
3			
4		WDE350LVQ Black Drop-in Ceramic	
5		WDE350LVS Stainless Drop-in Ceramic	
6		WDE350LVQ White Drop-in Ceramic	
7		WDE350LVQ Black Drop-in Ceramic	
8		WDE150LVS Stainless Drop-in Coil	
9		RY160LXTS Stainless Slide-in	
		WDE150LVB White Slide-in	
		WDE150LVQ Bisquit Slide-in	
		WDE150LVT Stainless Freestanding G	

Figure 8.19. Combo Box minimized after selection

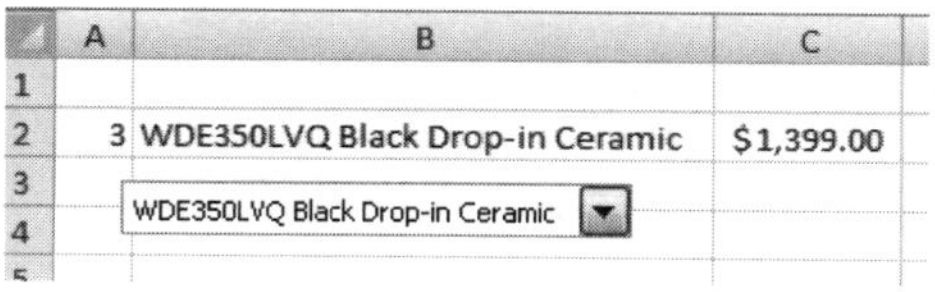

	A	B	C
1			
2	3	WDE350LVQ Black Drop-in Ceramic	$1,399.00
3			
4		WDE350LVQ Black Drop-in Ceramic	
5			

When you click off of the Combo Box, it collapses so that only the selected item is displayed. Figure 8.17 shows the completed Combo Box. Figures 8.18 and 8.19, respectively, show how the box expands to list options, and collapses as a selection is made.

Using Form controls enhances your worksheets. They are effective easy-to-create tools that make your spreadsheets more user friendly.

Automating Spreadsheets

9

Macros (mini computer programs) and User Forms that work in conjunction with them are powerful time-saving tools that enhance the functionality of your worksheets. A *User Form* incorporates *ActiveX* Controls, similar to Form tools. These controls use macros to automate many tasks, making your spreadsheets more user friendly. The Command Button is a Form tool you will use on a spreadsheet primarily to launch macros. This chapter teaches you how to

- create macros
- execute macros using "hot keys" attached to Command Buttons and other objects
- customize macro icons
- develop User Forms
- automate ActiveX Controls

You have already experienced Excel's® power in formulas, functions, and other features. Adding macros allows you to become even more efficient and effective so you can accomplish more work in less time.

Macros

A macro is a program you write or record that combines a series of commands so you can execute all of them with one or two clicks or keystrokes. Consider using a macro to automate a task that you must frequently repeat. You can simplify the following tasks with macros:

- entering routine information about your business (such as an address)
- formatting cells
- manipulating data

- creating reports
- performing mathematical computations

Excel® uses Visual Basic (VB) for Applications programming language to create macros. Learning VB requires time and expertise; however, Excel® has a user interface that allows you to record commands rather than having to write them manually in VB.

To create a macro, turn on the macro recorder, perform the steps to complete a task in the appropriate order, and turn the recorder off. Once you record a macro, you can attach it to an object, such as a button, drawing, picture, icon, or toolbar. Whenever a user clicks the object, the macro will run.

Creating a macro

Suppose Mike, of Quality Construction, wanted to create a macro that would automatically enter his company's name and address into a specific location on a spreadsheet. His company's name and address are:

Quality Construction

550 North Star Valley Drive

Dallas, Texas 75382-2877

After he creates the macro, he will be able to enter his company's name and address into a spreadsheet with one click. The macro will save time and prevent typos.

Select a Location

Before you create the macro, decide whether it should begin in the same location each time it is executed (absolute reference), or in the cell that is active when you execute the macro (relative reference). For a return address, you may want to enter the address in cell A1 in one workbook and B7 in another. In other words, if cell B7 is selected when the macro is executed, the three-row address would be placed in cells B7, B8, and B9.

- Click the **Developer** tab on the ribbon and click **Use Relative References** (fig. 9.1).

Figure 9.1. Relative reference

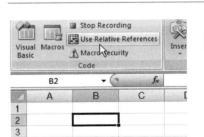

Name the macro

- Click the **Record macro icon** on the **Developer** tab of the ribbon (fig. 9.2) or at the bottom of the workbook (fig. 9.3). The Record macro box will pop up (fig. 9.4).

- Name the macro by typing *ReturnAddress* in the macro name box. (The name must be one word.) You also may enter a shortcut key that will execute the macro when you press **Ctrl** + the **hot key** (**A** in this case).

Figure 9.2. Record Macro button on the ribbon

Figure 9.3. Record Macro button on the status bar

Figure 9.4. Assigning a macro to the keyboard

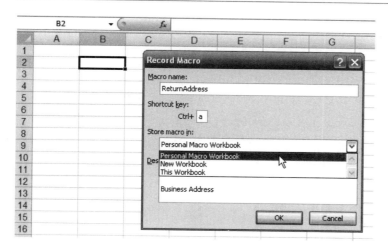

Store the Macro

- Select where you want Excel® to store the macro. Save macros to be used in multiple workbooks in the **Personal macro workbook**. The Personal macro workbook (*Personal.xlsb*) is hidden. Excel® creates it the first time you save a macro using the Personal macro workbook option.

- Select the *Personal.xlsb* file and click **OK**.

You will see the Personal macro workbook (*Personal.xlsb*) tab open for editing. Once you have created it, you can open this file as follows:

- Click the **View** tab on the ribbon.
- Click **Unhide** in the Window group.

The **Store macro in** option allows you to store macros in a new workbook (choose the **New workbook** option) or in the current workbook (choose **This workbook** option). Macros stored using these options will be available only with the workbook they are assigned to. They allow users to execute the macros even if the workbook is loaded on a different computer from the one where you created the macro.

Describe the macro

The last option in the pop-up macro box, **Description**, allows you to enter details about the macro such as the date it was created.

- After you finish entering data in the **Record macro** pop-up box, click **OK** to begin recording the macro. All clicks or keystrokes will be recorded until you turn off the recorder.

Type the macro

Now you are ready to type. Do not move the cursor from its location.

- Type the company name (fig. 9.5).
- Arrow down to the row below and type the street address.
- Arrow down and type the city, state, and zip code.
- Arrow down one more time.
- Click the **Stop Recording** icon on the **Developer** tab or at the bottom of the worksheet.

To see the VB code the macro recorder created, launch the VB editor (fig. 9.6).

- Click the **VB icon** in the **Developer** tab or press the **Alt + F11**.
- Look in *Module1* of the *Personal.xlsb* workbook.

Figure 9.5. Recording a macro

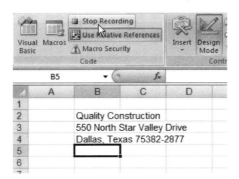

Figure 9.6. Return address macro in Visual Basic

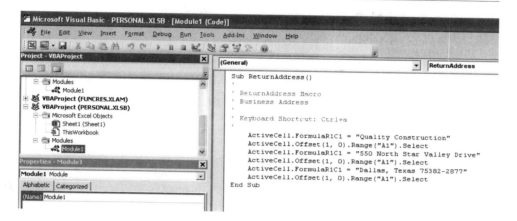

If other macros are stored on the Personal macro workbook, the *ReturnAddress* macro may be under another module (*Module2*, *Module3*, etc.). If you do not see the *Personal.xlsb* workbook, you may have saved your macro in the current workbook (possibly named *Book1*). If this is the case, you will need to recreate the macro, making sure that you select the Personal macro workbook option when you create the macro.

Notice the macro (subroutine) name listed at the top of the code. The information you typed in the Description box when you created the macro is *commented out*: it is preceded by a single quote. This means it is only informational and will not run when the macro is executed. The keyboard shortcut is also documented (**Ctrl + A**). Next, you see the VB code that executes when the macro is launched. The code states, "The active cell gets (=) whatever is in the quotes. Move down one row and zero columns and select that cell. Repeat the process 3 times."

- To test the macro, click a cell in the workbook and press the hot key, **Ctrl + A** (fig. 9.7).

If you want to launch the macro by clicking an object, such as a shape or picture on a worksheet, rather than a hot key

Figure 9.7. Return address typed (top) and executed as macro (bottom)

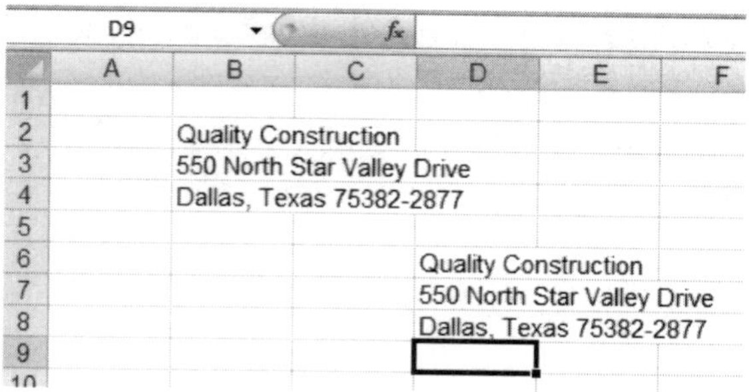

- Click the **Shapes** menu (fig. 9.8) on the **Insert** tab and drag the shape to where you want it on the worksheet.
- Resize the shape as desired.

You assign the macro to this object as follows:

- Right click the object.
- Click the **Assign macro** option from the list (fig. 9.9).
- Choose the macro to assign to the object.
- Click **OK.**
- Click off of the shape (fig. 9.10).

You also can assign hyperlinks to objects.

Figure 9.8. Shapes menu

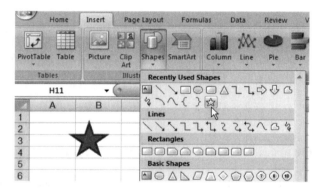

Figure 9.9. Assigning a macro to an object

Figure 9.10. Selecting the macro to attach to the object

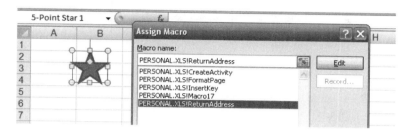

When you move the cursor over the object, the cursor will change to a pointing hand (fig. 9.11). This indicates the object is "hot" and when you click it, the macro will execute beginning in the cell that was active when the object was clicked.

You also can create hot objects on the QAT to execute macros. To add these custom icons to the QAT

- Click the down arrow to the right of the QAT (fig. 9.12).
- Click **More Commands** from the menu.

Figure 9.11. Running a macro by clicking the object

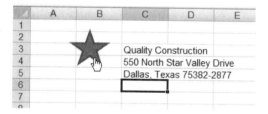

Figure 9.12. Adding a custom macro button

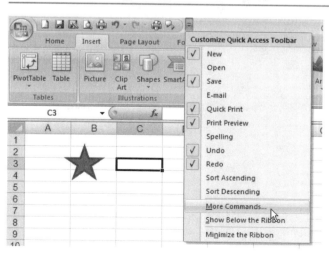

In the **Choose commands from** box, the **Popular Commands** will be displayed by default.

- From the down-arrow menu, choose **Macros** (fig. 9.13).
- From the list of macros displayed, choose the *ReturnAddress* macro, and the **Add** button (fig. 9.14).

To change the macro icon to another image

- Click the *ReturnAddress* macro displayed in the QAT list.
- Click the **Modify** button below the list (fig. 9.15).

Excel® displays a chart of premade icons to choose from.

- Choose the icon you prefer and click **OK**. ✎ Although earlier versions of Excel® allowed you to design custom icons on a bitmap editor, that feature does not exist in Excel® 2007.

If you need to correct a mistake or revise the macro (you might need to update an address, for example), you can easily edit VB code. Assume the company incorporated, so you want to change its title to include "Inc."

- Open the VB editor (click **Visual Basic** on the **Developer** tab or press **Alt + F11**).
- Place the cursor at the end of "Construction" and type "Inc." (fig. 9.16). *Don't* strike the **Enter** key after making the correction.
- To check whether the changes have been made, execute the macro through one of the three launch methods: (1) press the hot key (**Ctrl + A**), (2) click the object to which the code is assigned, or (3) click the associated icon on the QAT. You can see the results (fig. 9.17).

Figure 9.13. Customizing the Quick Access Toolbar

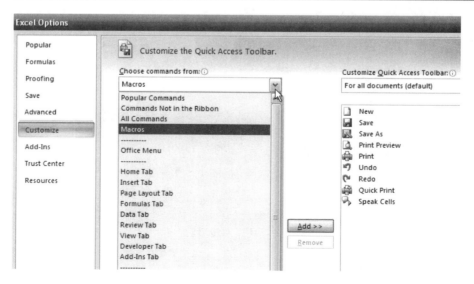

Figure 9.14. Adding the macro to the Quick Access Toolbar

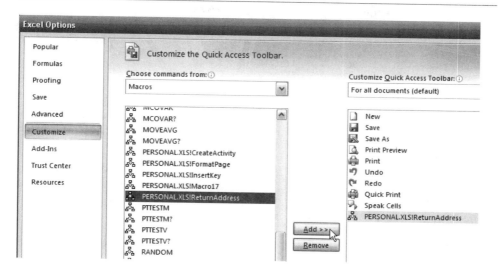

Figure 9.15. Modifying the macro icon

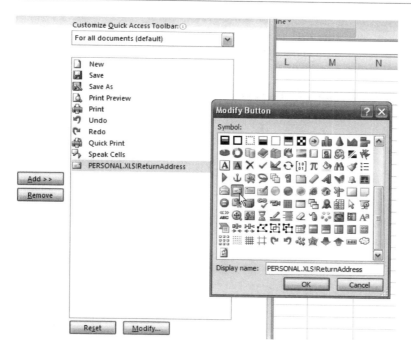

Figure 9.16. Editing the macro in Visual Basic

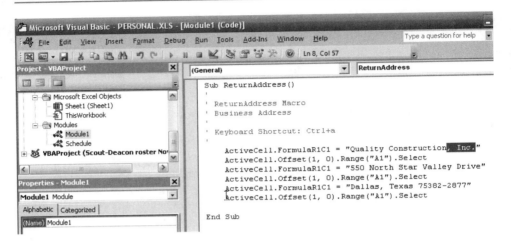

Figure 9.17. Final return address results

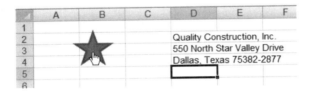

You can easily create and modify macros to use for a variety of purposes, saving time with repetitive tasks. But this only scratches the surface of how to use VB for Excel® to add power to your spreadsheets.

User Forms

A User Form is a pop-up box that can display information, enable optional tasks, and solicit information from users, among other functions. You can add objects such as Command Buttons, List Boxes, Labels, Spinners, or other items to User Forms. Each object can have code (procedures) attached to it. This code can be simple or complex.

In estimating, one of the most relevant applications for a User Form is creating a pop-up list of item descriptions in all of your detail sheets. These forms can replace cumbersome Data Validation. One limitation of Data Validation in picking information from a list is that each time you want to enter information into a specific cell you must select the cell, open the list, and finally, select the item. A faster and easier solution is to open the list and be able to keep it open while selecting items for multiple cells. After each selection, the cursor automatically moves down one cell.

Let's create a User Form that will allow users to pick items from a list by double clicking on them. Through this exercise, you will experience Excel®'s power to eliminate even more steps and save more time.

- Type a list of colors.
- Name it *ColorsList* as shown in figure 9.18 (*see also*, *Colors* worksheet in the Chapter 9 macro-enabled workbook).
- From the **Developer** tab on the ribbon, choose **Insert**.
- Click the **Button** tool on the **Form controls**.

Form controls are typically placed on a worksheet and ActiveX Controls are more often used with VB for Applications and User Forms.

- Click on the worksheet where you want to place the button.
- Drag down and resize the button as desired (fig. 9.19).
- Release the mouse button to display the button on the worksheet. Because Excel® anticipates that you will assign a macro to the button, it opens the **Assign macro** pop-up menu.
- Click **New** to create a new macro.

Excel® opens the VB editor and enters the first and last lines of the macro code (fig. 9.20). Each button you add is called a *CommandButton* in VB and the code assigns a number to each one. When a user clicks *Command Button1*, you want Excel® to display the list of colors.

Figure 9.18. Creating a macro button

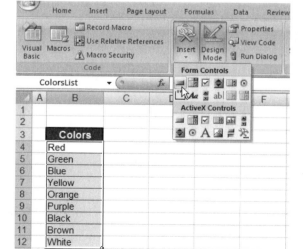

Figure 9.19. Placing a macro button on the worksheet

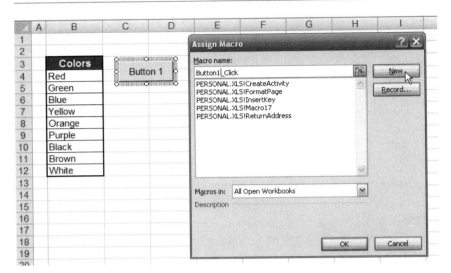

- Type **Userform1.show** on the blank line between the first and last lines of code VB has inserted for you.
- Click **Insert/Userform**.

Figure 9.20. Inserting a User Form

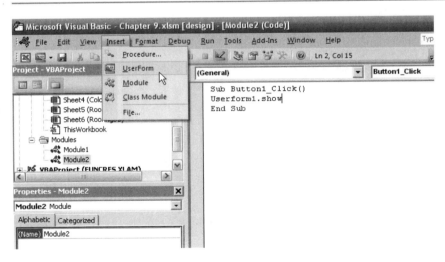

Excel® creates a blank User Form (*Userform1*) and displays the ActiveX toolbox with controls you can place on the User Form (fig. 9.21).

- Click the **List Box** control and place it on the left side of the User Form.
- Place two buttons on the right side of the **User Form.** You may resize the **User Form** and the **List Box** to fit the data you wish to display.

Figure 9.21. Designing the User Form

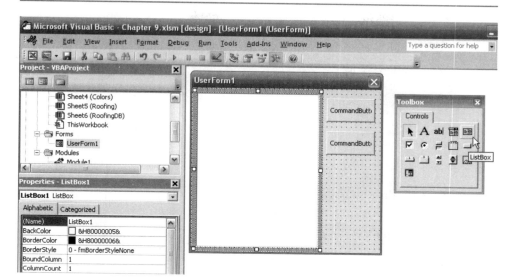

Notice the properties window in the lower left area of the screen. You can set properties on the User Form and its controls to manipulate the objects. The first property of any object is its name. Excel® enters the default names (*Userform1, Listbox1*, etc.); however, unique and more descriptive names can be assigned to any of these objects. So the user will understand the purpose of the form, change its generic name to *Pick a Color* or enter another appropriate description or instruction.

- Click the **down arrow** next to the User Form name in the **Properties** window and select *ListBox1* from the drop-down menu.

- Using the scroll bar in the Properties window, scroll to down to the *Row-Source* property for *ListBox1* (fig. 9.22).

- Enter the named range, *ColorsList*, which is the source of the list to populate ListBox1. When you enter *ColorsList* correctly as the *RowSource* property, Excel® will display the list of colors in *ListBox1*.

Now change the caption for both command buttons. You want the user to click the first button to enter the chosen color into the spreadsheet.

- Click *CommandButton1* and change the caption property to **OK** (fig. 9.23).

After they finish entering items into the spreadsheet, you want users to be able to close the User Form.

- To change the caption property of the second button to *Close,* select *CommandButton2*.

- Scroll down to **Caption** in the **Properties** window and type *Close* on the line next to Caption (fig. 9.24).

Now, the user will recognize what the buttons are supposed to do. However, before these buttons will operate as intended, you must attach code to them.

Figure 9.22. Setting the rowsource property

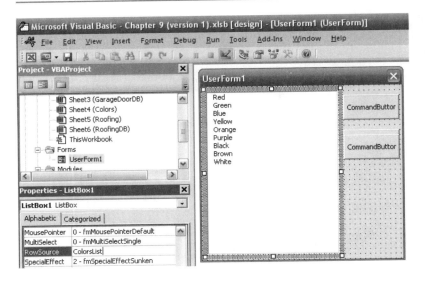

Figure 9.23. Changing the Command Button caption to "OK"

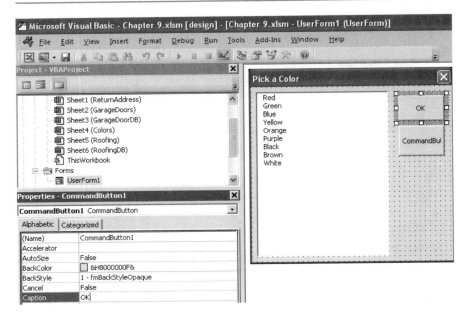

Figure 9.24. Changing the Command Button caption to "Close"

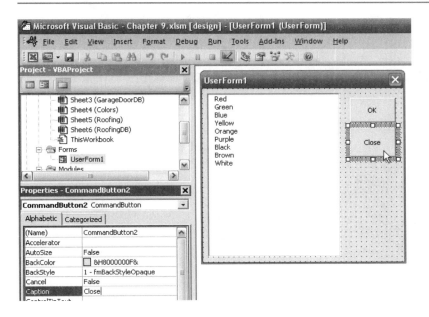

Go back to the spreadsheet to label the button *Add a Color*:

- Right click the button. A menu is displayed.
- Click **Edit Text** and enter *Add a Color*.
- Click off of the button to deselect it.
- Click the **Add-a-Color** button and Excel® will display *UserForm1* with the list of colors (fig. 9.26).

As noted previously, the objects on *UserForm1* are inoperable because they do not have any code attached yet. To make them operable, you need to go back into the VB editor to add code to them.

Figure 9.25. Editing the text of a Command Button

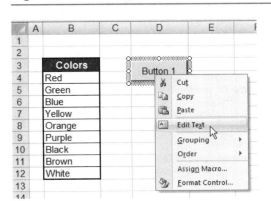

Figure 9.26. User Form in Excel®

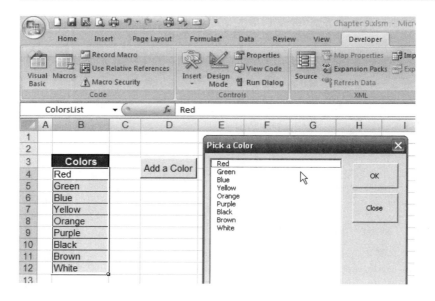

- When you activate the VB editor, *UserForm1* should still be displayed. If it is not, then click **Forms/UserForm1** (or the name you assigned to *UserForm1*) in the Project window on the left side of the screen (fig. 9.27).

- Double click the **OK** button (*CommandButton1*) and Excel® will display the sheet that holds the code assigned to *UserForm1*.

Again, VB has automatically created the first and last lines of the macro automatically. Now, you need to enter code that will place the color selected from the List Box into the active cell in the spreadsheet and then move the cursor down one cell so the user can select an additional color.

Figure 9.27. Displaying *Userform1*

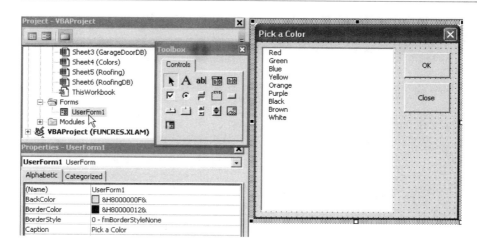

The cursor should be at the blank line between the first and last lines of VB code that Excel® inserted automatically. At that blank line, type the following two lines of code:

ActiveCell.Value=ListBox1.Value

ActiveCell.Offset(1,0).Select

The first line directs the program to place the value selected in *ListBox1* into the active cell. In other words, the value of the active cell gets "=" the value of *ListBox1*. The second line tells Excel® to move the cursor down 1 row and 0 columns and select the new cell.

Now, add a line of code to *CommandButton2*, the Close button, which will close UserForm1 from view.

- Double click the button (*CommandButton2*) on *UserForm1* to initiate the *CommandButton2_Click()* subroutine code editor.
- Under the Private subroutine for *CommandButton2_Click()*, type *UserForm1.Hide* (fig. 9.28).

After you enter the period (.), notice the help menu that VB offers to prompt and speed code entry. VB anticipates the possible instructions for *UserForm1*. You can scroll down the list provided or type the first letter of the property. For example, you can type *H*, click "hide," and press the space bar to complete the word, or you can just double click "hide."

To test the User Form, go to the worksheet and select a cell where you will begin entering colors.

- Click the **Add a Color** macro button to display *UserForm1* (fig. 9.29).
- Pick a color from the List Box and click **OK**. If you have programmed the macro correctly, Excel® will enter the color into the active cell and then move down one row.

Now apply what you have learned to creating a garage door detail sheet and an associated database. Name these worksheets, respectively, *GarageDoors* and

Figure 9.28. Adding code to Command Buttons

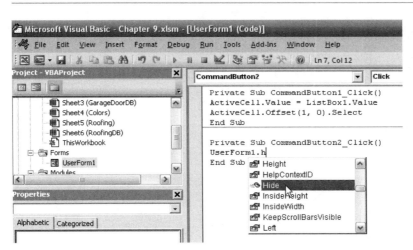

Figure 9.29. Adding colors

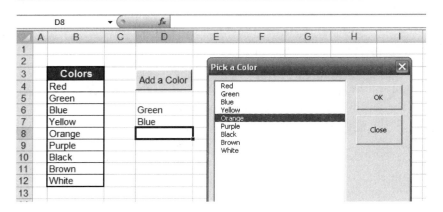

GarageDoorDB. The *GarageDoorDB* includes the named range *GarageDoorList.* It comprises the cells in the database with the garage door descriptions (A3:A28). The code you write will refer to this range.

You also need a roofing detail sheet and an associated database. Use the ones in the Chapter 9 workbook or create your own.

- Insert an "add item" button the garage door detail sheet. The **Assign macro** box pops up (fig. 9.30).

- Click **New** to open the VB editor. Note that Excel® automatically enters the initial code for the Button subroutine. The first subroutine on the page is *Button1_Click()*, the same code used for the choose-a-color button.

Figure 9.30. Assign macro pop-up box

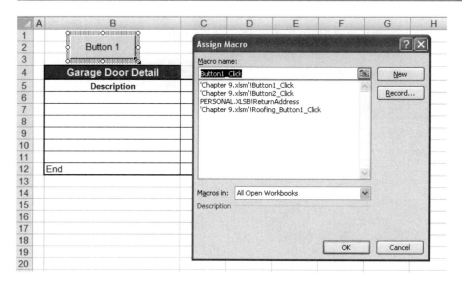

Enter the code for the second subroutine—for *Button2_Click()*. Excel® automatically writes the first and last lines of code for the button in VB. Set the User Form caption property to "Select Garage Door Items" and the List Box *Rowsource* property to *GarageDoorList* (fig. 9.31). Add the following code to the *Button2_Click()* subroutine:

UserForm1.Caption="Select Garage Door Items"

UserForm1.ListBox1.RowSource="GarageDoorList"

UserForm1.Show

Line one changes the *UserForm1* caption to read, "Select Garage Door Items." Line two sets the source of the *ListBox1* items to the list of items in the garage door database. The third line tells Excel® to display *UserForm1*.

There is only one User Form (UserForm1), but you can use this form to select items for two different detail sheets—colors and garage doors. You don't need to create a separate User Form for each purpose; you only need to change the properties to correspond to the specific detail sheet and its associated database (colors or garage doors).

To test the form

- Select the first cell in the **Description** list of the garage door detail sheet (fig. 9.32).

- Click the **Add Item** button. Excel® will display the User Form with the designated caption. The List Box will show the items from the description column of the garage door database.

- After selecting items from the List Box, click **OK**.

You can program the macro to position the cursor automatically before opening the User Form instead of having to position the cursor manually in the cell where you want to start adding item descriptions (fig. 9.33). Here's how:

Figure 9.31. Setting the User Form properties

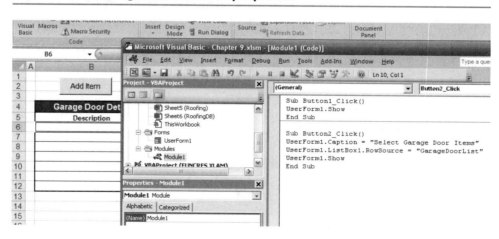

Figure 9.32. Testing the User Form

	A	B	C	D	E	F	G
1							
2		Add Item					
3							
4		Garage Door Detail					
5		Description	QTY	Unit	Unit Cost	Total Cost	
6		18x8 Non-Insulated	1	EA	735.74	735.74	
7							
8							
9							
10							
11							
12		End					
13							
14							
15							
16							

Select Garage Door Items

- 18x8 Insulated
- 18x8 Non-Insulated
- 18x7 Insulated
- 18x7 Non-Insulated
- 16x8 Insulated
- 16x8 Non-Insulated
- 16x7 Insulated
- 16x7 Non- Insulated
- 9x8 Insulated
- 9x8 Non-Insulated

OK

Close

- Name the range of cells in the description area of the garage door detail sheet (B6:B12) *GarageDoorInput*.
- Name the cells in the description area of the roofing detail sheet *RoofingInput*.
- Insert the following code into the first two lines of the subroutine, *Button2_Click()* (fig. 9.34):

Application.Goto Reference:="GarageDoorInput"

Selection.SpecialCells(xlCellTypeBlanks).Select

The first line of code selects the named range of cells, *GarageDoorInput*. An alternative equivalent line of code is **Range("GarageDoorInput").Select**. The second line of code selects the first blank cell in the selected range.

Figure 9.33. Adding code to place the cursor

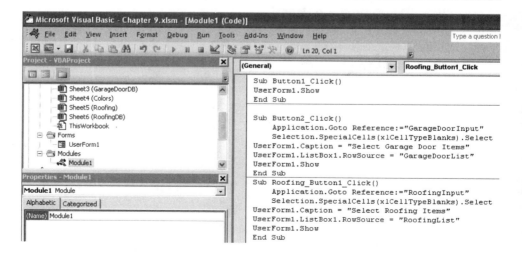

```
Sub Button1_Click()
UserForm1.Show
End Sub

Sub Button2_Click()
    Application.Goto Reference:="GarageDoorInput"
    Selection.SpecialCells(xlCellTypeBlanks).Select
UserForm1.Caption = "Select Garage Door Items"
UserForm1.ListBox1.RowSource = "GarageDoorList"
UserForm1.Show
End Sub
Sub Roofing_Button1_Click()
    Application.Goto Reference:="RoofingInput"
    Selection.SpecialCells(xlCellTypeBlanks).Select
UserForm1.Caption = "Select Roofing Items"
UserForm1.ListBox1.RowSource = "RoofingList"
UserForm1.Show
End Sub
```

Figure 9.34. Adding roofing items

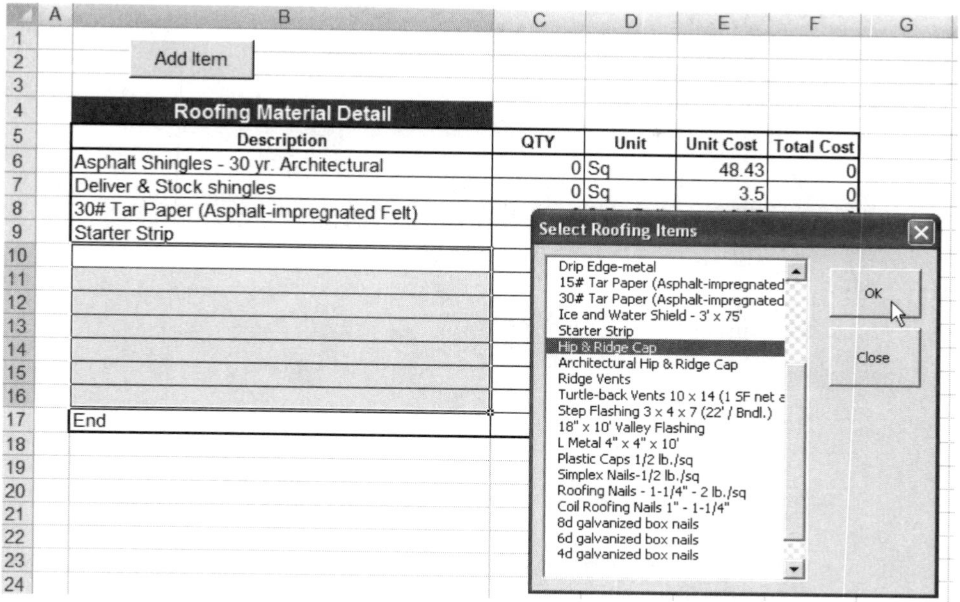

You can use the macro recorder to create the first two lines of code. For the first line

- Click **Find & Select** on the **Home** tab.
- Click **Go To.**
- Select the named range, *GarageDoorInput.*

For the second line

- Click **Find & Select** and **Go To Special.**
- Select **Blanks.**
- Click **OK.**

To add a Command Button to the roofing detail sheet

- Click **Insert** on the **Developer** tab.
- Select the **Button (Form Control).**
- Click and drag on the worksheet to where you want the button to be located.
- When the **Assign Macro** box pops up, click **New.**

Once again, the first and last lines of the code (*Sub Roofing_Button1_Click()* and *End Sub*) are entered into Module1.

- **Copy** the code from *Button 2_Click()* to *Roofing_Button1_Click().*
- **Edit** the input range, caption name, and *RowSource* identifier.
- Test the macro to ensure it correctly positions the cursor by executing the add-item macros for the garage door and roofing detail sheets (fig. 9.34).

Figure 9.35. Programming a User Form to respond to a double click

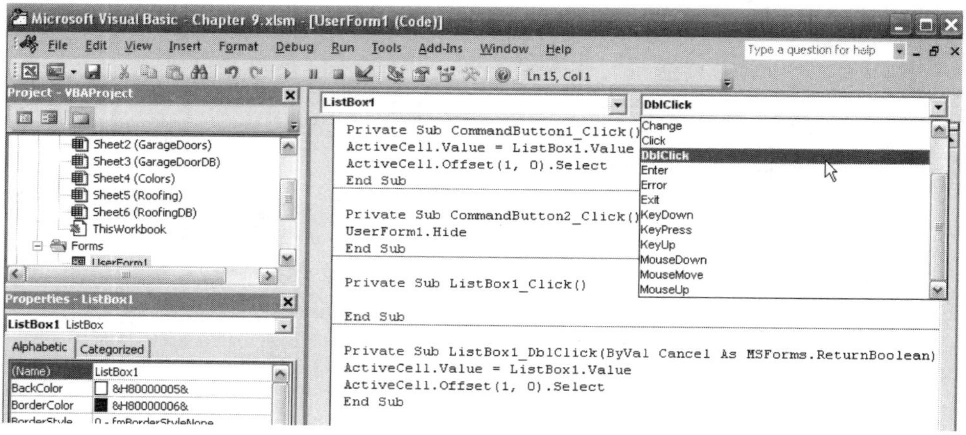

By copying and editing existing code, you can easily create Form Controls to apply to various phases of estimating.

Clicking OK each time you want Excel® to insert data into a worksheet can be cumbersome. You can accelerate the process by programming the User Form to respond to a double click instead.

- In the VB editor, display *UserForm1* and double click *ListBox1*. VB inserts the following code (fig. 9.35):

Private Sub ListBox1_Click()

End Sub

ListBox1 is a click event: If *UserForm1* is displayed in Excel® and the user clicks *ListBox1*, Excel® will execute the code between *Private Sub ListBox1_ Click()* and *End Sub*. To create a double-click event for *ListBox1*

- Click the **Event menu down arrow** to display the drop-down list at the top right of the editing window.

- Choose **DblClick**. Excel® creates a new subroutine.

- To enter the code that will run when a user double clicks *ListBox1*, **copy** the code from *CommandButton1_Click()* to the *ListBox1_DblClick* subroutine.

- Test the new capability of your User Form.

Ending a Detail Sheet

I often get these two questions from Excel® users: "When I have a long list of items to add to my detail sheet, how do I keep from adding these items below the bottom of the formatted area?" and "How do I automatically copy formulas as I add items to my estimating detail sheet?"

The VB code needs to know where the bottom of the detail entry area is in order to stop adding items. Indicate this by typing the word *End* in the cell (fig. 9.36).

Figure 9.36. Indicating the end of a User Form

	A	B	C	D	E	F
1						
2		Add Item				
3						
4		**Roofing Material Detail**				
5		Description	QTY	Unit	Unit Cost	Total Cost
6		Asphalt Shingles - 30 yr. Architectural	0	Sq	$ 48.43	$ -
7		Deliver & Stock shingles	0	Sq	$ 3.50	$ -
8		30# Tar Paper (Asphalt-impregnated Felt)	0	2-Sq Roll	$ 10.95	$ -
9		Starter Strip	0	Sq (T)	$ 38.00	$ -
10		Hip & Ridge Cap	0	Sq (T)	$ 38.00	$ -
11		Ridge Vents	0	Lf	$ 2.30	$ -
12		Step Flashing 3 x 4 x 7 (22' / Bndl.)	0	Bndl of 50	$ 12.95	$ -
13		L Metal 4" x 4" x 10'	0	Ea	$ 4.79	$ -
14		Plastic Caps 1/2 lb./sq	0	Pail	$ 34.95	$ -
15		Roofing Nails - 1-1/4" - 2 lb./sq				
16						
17		End				

You can change the text color to white so the word *End* will be invisible to users. Afterward, you can program VB to look for the word *End* If the program finds the word *End* it will insert a new row, copy the formulas from the previous row, and leave the description cell blank and ready for the next item to be entered. To execute these steps, you must include the following VB code in both the *CommandButton1_Click()* and the *ListBox1_DblClick(…)* subroutines (fig. 9.37).

```
Private Sub ListBox1_DblClick(ByVal Cancel As MSForms.ReturnBoolean)
Dim TempCell As String
Dim TempRow As Integer
ActiveCell.Value=ListBox1.Value
ActiveCell.Offset(1, 0).Select
If ActiveCell.Value="End" Then
        TempCell=ActiveCell.Address
        TempRow=ActiveCell.Row
        Rows(TempRow).Select
        Selection.Insert Shift:=xlDown
        Rows(TempRow − 1).Select
        Selection.Copy
        Rows(TempRow).Select
        ActiveSheet.Paste
        Range(TempCell).Select
        Range(TempCell).ClearContents
    End If
End Sub
```

In the program, *Dim* means dimension. It is a command for declaring a variable name and type.

Figure 9.37. Adding rows to the User Form

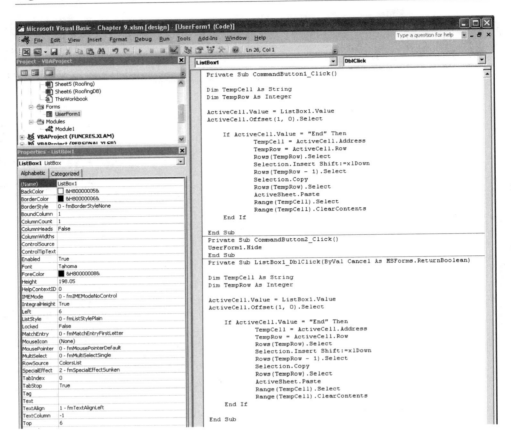

Test the code. Don't be afraid to try different settings and add your own controls to the User Form or your own code to the procedure modules. You will see the remarkable power of VB for Applications that comes with Microsoft Excel®.

Macros and User Forms are fascinating and powerful features of Excel® that deserve your time and attention. They will make your company more efficient in the long run. Customized spreadsheets will reduce your estimating time and you can use them for many other office tasks too. The effort you make now to implement these time-saving techniques in your business are as important as "sharpening your saw." In today's competitive environment, the tools that help you to become more effective and efficient also will make you more profitable.

Notes

1. The division numbers and titles are from *MasterFormat*™, published by The Construction Specifications Institute (CSI) and Construction Specifications Canada (CSC) and are used with permission from CSI, 2009, 99 Canal Center Plaza, Suite 300, Alexandria, VA 22314, 800-689-2900, 703-684-0300, http://www.csinet.org.

2. *Building Valuation Data.* International Code Council: Washington, DC.

3. *International Building Code*®. International Code Council: Washington, DC, 2009.

4. *1997 Uniform Building Code.* International Code Council: Washington, DC, 1997.

5. *The Cost of Doing Business Study*, 2010 edition. Washington, DC: Builder-Books, 2010.

Resources

Asdal, William. *Defensive Estimating: Protecting Your Profits*. Washington, DC: BuilderBooks, 2006.

Christofferson, Jay. *EstimatorPRO™ 5.2*. Washington, DC: BuilderBooks, 2010.

Dodge, Mark and Craig Stinson, *Microsoft® Office Excel®2007 Inside Out*. Redmond, WA: Microsoft Learning, 2007.

Frye, Curtis D. *Microsoft® Office Excel®2007 Step by Step*. Redmond, WA: Microsoft Learning, 2007.

Householder, Jerry with Emile Marchive III. *Estimating Home Construction Costs, Second Edition*. Washington, DC: BuilderBooks, 2006.

Walkenbach, John. *Microsoft Excel® 2007 Bible*. Indianapolis: Wiley Publishing Inc., 2007.

Walkenbach, John. *Excel® 2007 Power Programming with VBA*. Indianapolis: Wiley Publishing Inc., 2007.

Index

A

ActiveX Control(s), 121, 131–32
area, calculating, 12–13
   circle, formula for, 12–13
Autofill, vii, 18, 42
AutoSum, viii, 37–39, 48

B

Building Valuation Data, 90, 145
button, vii, viii, x, xi, xii, 2, 5–9, 19–21, 37,
   39, 42–43, 45, 75, 77, 108–9, 111–15,
   117, 121–23, 127–28, 130–44
   Close, 5, 137
   Help, 5, 55, 68
   Office, vii, 2, 5–7, 9, 108–9
   Option, 112–14, 123–26
   Redo, vii, 6
   Undo, vii, 6

C

cell contents, 6, 20
   copying, 20
   moving, 20
   formatting, 19, 40–42, 46, 70
   naming, vii, viii, 10, 17, 50–51, 62

range, vii, 9–11, 17–21, 37, 42, 50–51, 53,
   55, 65, 68–69, 83, 92–93, 97, 115, 117,
   133, 138, 140–143
circular reference, 107–8
column, resizing, 21
Combo Box, xi, 118–19
concrete, ix, 12, 15, 26, 28, 61, 79–88
cost breakdown summary sheet, v, 24–25, 29,
   35–46, 73–76
cubic yards (CY), calculating, 12, 61, 80,
   83–88
cursor, changing default movement of, vii,
   8–9

D

Data Validation, viii, ix, 50–53, 81, 87–88, 130
database, viii–x, 33, 47, 49–51, 53, 55, 57,
   61–63, 65, 67, 70, 72–73, 80–82, 84,
   86–92, 95, 112, 117–18, 137–39
detail sheet, viii–x, 31–33, 47–49, 51, 61–62,
   65–75, 80–81, 86–89, 93–96, 99, 112,
   137–43
direct cost(s), x, 24, 103–5
disbursement. *See* draw
draw, 26, 29, 36, 102, 114

E
errors, ix, 53–56, 66–69, 82–84, 107

F
fee(s), x, 24–25, 79, 90–96, 99–100, 103
   appraisal, 99
   impact, 79, 90
   inspection, 99
   closing, 99
   connection, x, 79, 90, 93–96
   origination, 99–100
   permit, x, 24, 79, 90–94
fill handle, ix, 18–19, 38–39, 42, 48, 52, 55,
   84–85
footing(s) ix–x, 23–26, 29–30, 85–89, 96
Form control(s), v, 109, 111–19, 131,
   141–42
formula(s), vii, ix–x, 6, 10–19, 37, 40–42, 48,
   50–57, 60–69, 74, 81–85, 88–96, 102,
   104, 110–114, 117
   bar, 6, 10, 14, 37–39, 50, 55, 111
   copying, 38, 42, 48, 55, 57, 81, 84, 93, 96,
    111, 142–43
   naming, 12–13
   view, 62
function(s), v, viii, ix, xiii, 2, 5–6, 12, 19, 24,
   39–40, 43, 45, 47, 49, 53, 55–57, 61,
   65–69, 73, 79, 81, 83–85, 87, 89–93, 95,
   97, 113, 121, 130
   IF, ix, 56, 66, 113
   ISBLANK, 83
   ISERROR, 83
   MATCH, 67–69, 83–84, 91–95
   NOW, viii, 39
   Paste Special, 43
   ROUNDUP, 12, 61
   Sort, 45
   SUMIF, 65
   Transpose, 45
   VLOOKUP, 53, 94–95
Function Wizard, viii, 40, 55, 57

G
gross sales, 103, 105

H
hardware, computer, 2–3
hyperlink, 10, 75–76

I
interest, loan, calculating, 11–12, 99–100, 102

L
labor, 47, 57, 62, 67–70, 73, 81–88, 91, 103
link. *See* hyperlink
List Box, x, xi, 116–18, 132, 136–37, 139
littorals, 10

M
macro, xi, 2, 121–32, 136–39, 141
margin, v, x, 25, 30, 33, 99–108
markup, x, 105–106
multiply, viii, 10, 12, 15, 23, 32–33, 43–44, 48,
   60–61, 64, 80, 83, 85, 90, 92, 104, 111

N
Name box, vii, 10–11, 13, 15–16, 50, 89, 109,
   123
NAHB Chart of Accounts, 25, 36, 77
NOW function, viii, 39

O
operator, 10, 17, 43
overhead, 2, 24–25, 99, 103–5, 107–8

P
Paste Special, 43
Personal macro workbook, 123–25
profit, v, xiii, 1–2, 25, 30, 33, 35, 99–108
protecting a worksheet, ix, 70–71
psi, 80, 84, 86–87

Q
quantities, material, calculating, 2, 23, 31–33,
   47, 49–50, 57–58, 62, 65, 70, 79–80,
   88–90, 99, 114
Quick Access Toolbar (QAT), 6, 127–28

R
radius, 12–14
Random Access Memory (RAM), 2
rebar, x, 15, 26, 79, 85–89, 96
roofing, viii, ix, xii, 25, 27, 30–33, 43, 47–58,
   61–70, 73–76, 138, 140–41
   asphalt-impregnated felt, 63
   hip factor, 64
   Ice & Water Shield®, 63–64

ridge cap, 62, 65
shingles, 32, 52, 55, 58–67, 115
slope factor, ix, 59–60, 64
starter strip, 59–65
ROUNDUP function, 12, 61

S
schedule, 24
scroll bar, x, 114–15, 133
S curve, x, 100–101
sheathing, 60
Sheet tab menu, vii, 8
software, 1, 3, 73, 87
Sort, 45
spinner, x, 115–16
spreadsheet, viii, 2, 20, 24, 31–32, 41–42, 47,
    49, 62, 66, 70, 79, 105, 121–22, 133,
    135–36
Status bar, xi, 6, 123
summary (cap) sheet, v, 2, 24–25, 29, 35–46,
    73–76

T
tab, 6, 37, 50, 67, 74
    Control, 111, 113, 115
    Data, viii, 45, 51–52, 81, 87
    Developer, 109, 110, 122–23, 128, 131, 141
    Formulas, 11, 13–17, 37, 108
    Home, vii, 12, 19–20, 37, 70, 75, 141
    Insert, 126
    Insert Worksheet, 7, 75
    Number, 41–42

Personal macro workbook, 124
Protection, 70
Review, 97
Settings, 87
Sheet, 8
Show Developer, x
Transpose, 45
tab-and-ribbon, vii, 5
table, x, xii, 10, 17, 33, 50, 53, 55, 90–92,
    100, 105–6
tax, calculating, 15–17, 85, 93
technology, xv, 1
Trace Dependents, 108
Transpose, 45

U
User Form, xi, xii, 121, 130–44

V
variable, 10, 13, 143
variance, cost, viii, 29–30, 38–39
Visual Basic (VB), xi, 122, 125, 128, 130

W
waste factor, 60, 83, 88
Windows operating system, 2
workbook(s)
    Excel®, vii, 2, 6–7
    on CD, 73, 80, 85, 87, 105, 110, 131
worksheet
    moving and renaming, 7
    tab, 37, 50